BITE-SIZED CALCULATIONS IN CHEMISTRY

Mass, Mole and Stoichiometry

John Lambert

TowerGate Press

The characters and events portrayed in this book are fictitious. Any similarity to real persons, living or dead, is coincidental and not intended by the author.

ISBN: 9798332769610

Cover design by: The Author
Printed in the United States of America

To my wife and children

CONTENTS

PREFACE

Chemistry is a subject that cannot be fully appreciated without a full mastery of the diverse numerical problems that are encountered in almost all aspects of the subject. Unfortunately, experience has shown that it is this aspect of chemistry students fear the most. This series of book, Bite-Sized Chemistry Calculations, is aimed at helping students overcome the challenges associated with tackling the various types of calculations encountered in different aspects of chemistry, focusing on a few topics at a time to facilitate comprehension.

As an experienced chemistry educator who understands the many challenges faced by students, I have carefully tailored each book of the series to fully meet the needs of students at all levels, especially those taking college level general chemistry courses as well as those following various O-level curricula worldwide. This part of the series explores the different types of calculations found in formulae, the mole and stoichiometry. It covers the determination of formula of ionic compounds, relative formula masses, mass and percent compositions of compounds, all aspects of mole calculations, empirical and molecular formulae, balancing of chemical equations, calculations based on chemical equations, limiting reagents and percent yield.

The series is packed with many salient features that are meant to facilitate both teaching and learning. Some of these include helpful explanations, many examples, alternative ways to solve problems, plenty of practice questions and complete answers. With this book, you will be well prepared for your exams and boost your performance.

John Lambert
July 7, 2024

CHAPTER 1: WRITING THE FORMULAE OF IONIC COMPOUNDS

A chemical formula, or formula unit, is the short way of writing the name of a chemical substance, using only the chemical symbols of its constituent elements. The chemical formula of water, for example, is H_2O. Chemical formulae are very important in chemistry; for example, chemical formulae are required to perform mass calculations and write chemical equations.

Unlike covalent compounds, ionic or electrovalent compounds are not molecules. They are made up of positively and negatively charged ions, i.e., electrically charged atoms or group of atoms that are held together by electrostatic force of attraction. The positively charged ions are called cations, while the negatively charged ions are called anions.

Writing the chemical formula of an ionic compound is quite straightforward as long, as its constituent ions or elements can be easily identified from its IUPAC name. For example, the compound magnesium chloride indicates the presence of magnesium ions, Mg^{2+}, and chloride, Cl^-, ion. It is thus an important skill in chemistry to be able to write the ions of elements correctly.

The formula of an ionic compound is written in such a way that the compound possesses no net electric charge. The procedure is as follows.

1. Starting with the positively charged ion, write the constituent ions and radicals of the compound.

2. Balance the charges on the two ions by interchanging the valencies (the charges without their signs) of the elements and radicals in the compound, and add them to the ions as subscripts. Note that subscripts of '1' are not written but assumed.

3. Polyatomic ions, also called radicals in old literature, such as OH- and NO^{3-}, must be enclosed in parentheses before adding subscripts.

4. Finally, write the formula by dropping the charges on the ions.

Example 1

Deduce the chemical formula of the compound formed between sodium ions, Na^+, and chloride ions, Cl^-.

Solution

We begin by writing the two ions next to each other, starting with the anion, i.e., Na^+Cl^-. Each ion has a single charge. Thus, the net charge is zero, i.e. net charge = 1 – 1 = 0. Since the charges are balanced, then the required chemical formula is NaCl.

Example 2

Deduce the chemical formula of calcium chloride.

Solution

As usual, the very first step is to identify the constituent ions of the given compound. The name of the compound indicates that the ions present are magnesium ion, Ca^{2+}, and chloride ion, Cl^-. Thus, we should now write $Ca^{2+}Cl^-$. We must now balance the charges by adding a subscript of 2 (the valency of Ca) to Cl^-, i.e. $Ca^{2+}Cl^-_2$. Finally, we can now drop the charges to obtain the formula $CaCl_2$. Note that the subscript of Ca is 1 and, as stated earlier, subscripts of '1' are not written but assumed.

Example 3

Deduce the formula of aluminium sulfate.

Solution

The compound contains aluminium ion, Al^{3+}, and sulfate, SO_4^{2-}, a polyatomic ion; hence, we write $Al^{3+}SO_4^{2-}$. We must now balance the charges by adding a subscript of 2 (the valency of SO_4^{2-}) to Al^{3+}, and a subscript of 3 (the valency of Al) to SO_4^{2-}. We must, of course, remember to enclose SO_4^{2-} in parentheses before adding the subscript, i.e., $Al_2^{3+}(SO_4^{2-})_3$. Thus, the required formula is $Al_2(SO_4)_3$.

Example 4

Write the chemical formula of iron(II) hydroxide.

Solution

Iron is a typical example of multivalent metals, i.e., metals that form multiple ions. The figure, written in in parenthesis indicates that the ion of iron in the compound is Fe^{2+}. The anion is hydroxide, OH^- (a polyatomic ion); hence, we write $Fe^{2+}OH^-$. We must balance the charges by enclosing OH^- in parentheses and adding a subscript of 2 (the valency of Fe in the compound), i.e., $Fe^{2+}(OH^-)_2$; hence, the formula of iron(II) hydroxide is $Fe(OH)_2$.

Practice Problems

Deduce the formulae of the following compounds.

(1) Silver chloride

(2) Calcium hydroxide

(3) Ammonium sulfate

(4) Iron(II) chloride

(5) Magnesium nitrate

CHAPTER 2: FORMULA MASSES

Relative atomic mass

The masses of atoms are too small to be of any practical use. Instead, they are usually expressed relative to some standard. This standard of comparison is carbon-12, the stable isotope of carbon, which is assigned a value of 12 units.

The relative atomic mass A_r of an element is the number of times the average mass of one atom of the element is heavier than one-twelfth the mass of an atom of carbon-12. This is given as follows.

$$A_r = \frac{\text{Average mass of an atom of an element}}{\frac{1}{12}\text{mass of an atom of carbon} - 12}$$

Relative atomic mass has no units since it's a ratio of two masses. The approximate relative atomic masses of the first twenty elements are given in Table 1.1.Table 1.1 The approximate relative atomi masses of the first twenty elements.

Table 2.1: The Approximate Relative Atomic Masses of the First Twenty Elements

Atomic number	Element	A_r
1	Hydrogen	1
2	Helium	4
3	Lithium	7
4	Beryllium	9
5	Boron	11
6	Carbon	12
7	Nitrogen	14
8	Oxygen	16
9	Fluorine	19
10	Neon	20
11	Sodium	23
12	Magnesium	24
13	Aluminium	27
14	Silicon	28
15	Phosphorus	31
16	Sulphur	32
17	Chlorine	35.5
18	Argon	40
19	Potassium	39
20	Calcium	40

Relative formula

The relative formula mass, M_r, of a substance is the number of times the average mass of one mole or molecule of that substance is as heavy as one-twelfth the mass of an atom of carbon-12. This is is given by

$$M_r = \frac{\text{Average mass of one mole or molecule of a substance}}{\frac{1}{12}\text{mass of an atom of carbon} - 12}$$

Relative formula mass has no units since it's a ratio of two masses.

The relative formula mass of a covalent compound or molecule is also called relative molecular mass. The relative formula mass of a substance is obtained by summing up the masses of the elements in its chemical formula or formula unit. The total mass of an element in a chemical formula is obtained by multiplying its relative atomic mass by the subscript of its symbol in the formula.

Formula mass

The formula or molar mass, *M*, of a substance is the mass of one mole of that substance. It is expressed in g mol^{-1} or g/mol (gramme per mole). The formula mass of a compound, or polyatomic ion, is its relative formula mass expressed in g mol^{-1}, while that of atoms of an element, or a simple ion, is its relative atomic mass expressed in g mol^{-1}. The formula mass of a covalent compound is also called molecular mass.

Example 1

Determine the relative molecular mass of oxygen molecule, O_2.

(O = 16)

Solution

Since there are two atoms in a molecule of oxygen, then

$$M_r = 16 \times 2 = 32$$

Example 2

Determine the relative formula mass of sodium, Na.

(Na = 23)

Solution

The relative formula mass of a metal is simply its relative atomic mass.

$$\therefore \quad M_r = 23$$

Example 3

What is the molar mass of chlorine atom?

(Cl = 35.5)

Solution

The molar mass of an atom of an element is its relative atomic mass expressed in g mol^{-1}. Thus,

$$M = 35.5 \text{ g mol}^{-1}$$

Example 4

Determine the molar mass of chlorine, Cl_2.

(Cl = 35.5)

Solution

As the formula suggests, chlorine, like oxygen, is a diatomic molecule.

$$\therefore \quad M = (35.5 \times 2) \text{ g mol}^{-1}$$
$$= 71 \text{ g mol}^{-1}$$

Example 5

Determine the molar mass of calcium oxide, CaO.

(O = 16, Ca = 40)

Solution

The following information is obtained from the formula.

Element	Number of atoms
Ca	1
O	1

$\therefore \quad M = \{(40 \times 1) + (16 \times 1)\} \text{ g mol}^{-1}$
$= 56 \text{ g mol}^{-1}$

Example 6

Determine the molar mass of water, H_2O.

(H = 1, O = 16)

Solution

The following information can be obtained from the formula.

Element	Number of atoms
H	1
O	2

$\therefore \quad M = \{(1 \times 2) + (16 \times 1)\} \text{ g mol}^{-1} = 18 \text{ g mol}^{-1}$

Example 7

Determine the molar mass of zinc hydroxide, $Zn(OH)_2$.

(H = 1, O = 16, Zn = 65)

Solution

We can obtain the following information from the formula

Element	Number of atoms
Zn	1
O	2
H	2

$\therefore \quad M = \{(65 \times 1) + (16 \times 2) + (1 \times 2)\} \text{ g mol}^{-1}$
$= 99 \text{ g mol}^{-1}$

Example 8

Determine the molar mass of iron(III) sulfate, $Fe_2(SO_4)_3$.

(O = 16, S = 32, Fe = 56)

Solution

We can obtain the following information from the formula.

Element	Number of atoms
Fe	2
S	3
O	12

$\therefore \quad M = \{(56 \times 2) + (32 \times 3) + (16 \times 12)\}\ \text{g mol}^{-1}$

$= 400\ \text{g mol}^{-1}$

Example 9

Determine the molar mass of cobalt(II) chloride hexahydrate, $CoCl_2.6H_2O$.

(H = 1, O = 16, Cl = 35.5, Co = 59)

Solution

We can obtain the following information from the formula.

Element	Number of atoms
Co	1
Cl	2
H	12
O	6

$\therefore \quad M = \{(59 \times 1) + (35.5 \times 2) + (1 \times 12) + (16 \times 6)\}\ \text{g mol}^{-1}$

$= 238\ \text{g mol}^{-1}$

Example 10

Determine the molar mass of an alkane that is represented as C_xH_{20}.

$$(H = 1, C = 12)$$

Solution

The very first step is to obtain the molecular formula of the given alkane. To do this we have to write the general formula of alkanes–C_nH_{2n+2}–and, then, equate corresponding subscripts as shown below.

$$n = x \quad (1)$$

$$2n + 2 = 20 \quad (2)$$

From Equation (2) we have

$$2n = 18$$

$$n = \frac{18}{2} = 9$$

From Equation (2) we know that n = x = 9. Inserting this in the above formula gives C_9H_{20}, from which we can obtain the following information.

Element	Number of atoms
C	9
H	20

$$\therefore \quad M = \{(12 \times 9) + (1 \times 20)\} \text{ g mol}^{-1}$$

$$= 128 \text{ g mol}^{-1}$$

Practice Problems

Determine the molar mass of each of the following compounds.

1. CO
2. CaO
3. NH3
4. N2O

5. $CaCl_2$
6. $CaCO_3$
7. H_2CO_3
8. $Ca(OH)_2$
9. NH_4NO_3
10. $FeSO_4.7H_2O$

(H = 1, C = 12, N = 14, O = 16, S = 32, Cl = 35.5, Ca = 40, Fe = 56)

CHAPTER 3: MASS PERCENT COMPOSITIONS

Mass composition

The mass composition of a substance gives the contribution of each of its constituent elements towards its overall mass. The mass of the element A in a substance with the formula A_xB_y *is* given by

$$m_A = \frac{xA_{rA}}{M_r} \times m$$

where

m_A = mass of A

A_{rA} = relative atomic mass of element A

M_r = relative molecular mass of the substance A_xB_y

m = mass of the substance A_xB_y

Mass percent composition

The mass percentage of the element A in a substance with the formula A_xB_y is given by

$$m_A = \frac{xA_{rA}}{M_r} \times 100\%$$

where

m_A = mass percent composition of A

A_{rA} = relative atomic mass of element A

M_r = relative molecular mass of the substance A_xB_y

The percent composition of an element in a pure substance is constant for any pure sample of the substance, regardless of the amount under consideration.

Example 1

Calculate the mass composition of 5.50 g of C_2H_4.

(H = 1, C = 12)

Solution

$$m_A = \frac{xA_{rA}}{M_r} \times 100\%$$

$$M_r = (12 \times 2) + (1 \times 4) = 28$$
$$m = 5.50\ g$$
$$m_C = ?$$
$$m_H = ?$$

Substituting we have

$$m_C = \frac{12 \times 2}{28} \times 5.50\ g = 4.71\ g$$

$$m_H = \frac{4 \times 1}{28} \times 5.50\ g = 0.79\ g$$

Example 2

The combustion of a hydrocarbon sample produces 2.2g of carbon dioxide and 1.8 g of water. Determine the masses of carbon and hydrogen in the original sample.

(H = 1, C = 12, O = 16)

Solution

We will apply the relation

$$m_A = \frac{xA_{rA}}{M_r} \times 100\%$$

The mass of carbon in the original hydrocarbon sample is equivalent to its mass in carbon dioxide, CO_2.

$M_r = (12 \times 1) + (16 \times 2) = 44$

$m = 2.2$ g

$m_C = ?$

Substituting we have

$$m_C = \frac{12 \times 1}{44} \times 2.2 \text{ g} = 0.60 \text{ g}$$

Similarly, the mass of hydrogen in the original hydrocarbon sample is equivalent to its mass in water, H_2O. Thus,

$M_r = (1 \times 2) + (16 \times 1) = 18$

$m = 1.8$ g

$m_H = ?$

Substituting we have

$$m_H = \frac{1 \times 2}{18} \times 1.8 \text{ g} = 0.20 \text{ g}$$

Example 3

The combustion of 1.2 g of an organic compound containing carbon, hydrogen and oxygen yields 1.3 g of carbon dioxide and 0.54 g of water. Determine the mass composition of the sample.

(H = 1, C = 12, O = 16)

Solution

$$m_A = \frac{xA_{rA}}{M_r} \times 100\%$$

For carbon we have

$$M_r = (12 \times 1) + (16 \times 2) = 44$$

$$m = 1.3 \text{ g}$$

$$m_c = ?$$

We now substitute to obtain

$$m_C = \frac{12 \times 1}{44} \times 1.3 \text{ g} = 0.35 \text{ g}$$

For hydrogen we have

$$M_r = (1 \times 2) + 16 = 18$$

$$m = 0.54 \text{ g}$$

$$m_H = ?$$

Substituting we have

$$m_H = \frac{1 \times 2}{18} \times 0.54 \text{ g} = 0.060 \text{ g}$$

The mass of oxygen is obtained by deduction.

$$m_o = 1.\,2 \text{ g} - (0.35 \text{ g} + 0.060 \text{ g}) = 0.79 \text{ g}$$

Example 4

Determine the mass percent composition of copper(II) oxide, CuO.

(O = 16, Cu = 64)

Solution

We will apply the relation

$$m_A = \frac{xA_{rA}}{M_r} \times 100\%$$

$$M_r = 64 + 16 = 80$$

$$m_{Cu} = ?$$

$$m_o = ?$$

Substituting we have

$$m_{Cu} = \frac{64 \times 1}{80} \times 100\% = 80\%$$

$$m_{O} = \frac{16 \times 1}{18} \times 100\% = 20\%$$

Note that the values must all sum up to 100%:

$$80\% + 20\% = 100\%$$

Example 5

Calculate the percentage by mass of each element in calcium nitrate, $Ca(NO_3)_2$.

(N = 14, O = 16, Ca = 40)

Solution

$$m_A = \frac{xA_{rA}}{M_r} \times 100\%$$

$M_r = (40 \times 1) + (14 \times 2) + (16 \times 6) = 164$

$m_{Ca} = ?$

$m_N = ?$

$m_O = ?$

Substituting we have

$$m_{Ca} = \frac{40 \times 1}{164} \times 100\% = 24.4\%$$

$$m_{N} = \frac{14 \times 2}{164} \times 100\% = 17.1\%$$

$$m_{O} = \frac{16 \times 6}{164} \times 10\% = 58.5\%$$

All values, as usual, must sum up to 100%:

$$24.4\% + 17.1\% + 58.5\% = 100\%$$

Practice Problems

1. Determine the mass of each element in 1.5 g of carbon monoxide, CO.
2. A hydrocarbon sample was burnt completely to produce 15.7 g and 6.4 g of carbon dioxide and water, respectively. Determine the masses of carbon and hydrogen in the original sample.
3. 5.0 g of a sample of a compound containing carbon, hydrogen and oxygen was burnt completely to produce 9.57 g and 5.85 g of carbon dioxide and water, respectively. Determine the mass composition of the original sample.
4. Determine the mass percentage compositions of the following compounds.
   (a) NH_3
   (b) CuO
   (c) Cu_2O
   (d) $CaCO_3$
   (e) H_2SO_4
   (f) $Ca(OH)_2$
   (g) $(NH_4)_2SO_4$
   (H = 1, C = 12, N = 14N = 14, O = 16, S = 32, Ca = 40, Cu = 63.5)

CHAPTER 4: THE MOLE AND MASS

What is the mole?

The mole, *n*, with the short form mol, is the SI unit for measuring the amounts of substances. One mole of a substance is defined as the amount of that substance containing exactly the same number of particles or elemental entities–atoms, ions or molecules–as exactly 12 g of carbon-12. In other words, a mole corresponds to exactly 12 g of carbon-12.

There are four different ways of calculating the number of moles of substances, depending on the nature of the substance under consideration. In this chapter, we will consider the relationship between the mole and mass.

The mole and mass

The number of moles of any substance is related to its mass, *m,* by

$$n = \frac{m}{M}$$

where *M* is molar mass.
We can see from the above relation that 1 mol of a substance corresponds to its molar mass. 1 mol of sodium chloride, for example, corresponds to 58.5 g of the salt.

Example 1

Determine the number of moles in 4.5 g of sodium carbonate, Na_2CO_3.

(C = 12, O = 16, Na = 23)

Solution

$$n = \frac{m}{M}$$

$$M = \{(23 \times 2) + (12 \times 1) + (16 \times 3)\}\ \text{g mol}^{-1}$$
$$= 106\ \text{g mol}^{-1}$$
$$m = 4.5\ \text{g}$$
$$n = ?$$

Substituting we have

$$n = \frac{4.5\ \text{g} \times 1\ \text{mol}}{106\ \text{g}} = 0.042\ \text{mol}$$

Example 2

A chemist weighed out 2.5 g of copper(II) sulfate, $CuSO_4$. How many moles of the compound did he weigh?

(O = 16, S = 32, Cu = 64)

Solution

We have to apply the relation

$$n = \frac{m}{M}$$

$$M = \{(64 \times 1) + (32 \times 1) + (16 \times 4)\}\ \text{g mol}^{-1}$$
$$= 160\ \text{g mol}^{-1}$$
$$m = 2.5\ \text{g}$$
$$n = ?$$

Substituting we have

$$n = \frac{2.5\ \text{g} \times 1\ \text{mol}}{160\ \text{g}} = 0.016\ \text{mol}$$

Example 3

Determine the mass of 0.15 mol of calcium oxide, CaO.
(O = 16, Ca = 40)

Solution

Recall that

$$n = \frac{m}{M}$$

From which

$$m = n \times M$$
$$M = \{(40 \times 1) + (16 \times 1)\} \text{ g mol}^{-1}$$
$$= 56 \text{ g mol}^{-1}$$
$$n = 0.15 \text{ mol}$$
$$m = ?$$

Substituting we have

$$m = 0.15 \text{ mol} \times 56 \text{ g mol}^{-1} = 8.4 \text{ g}$$

Example 4

0.0150 mol of a salt weighs 0.802 g. Determine the molar mass of the salt.

Solution

We know that

$$n = \frac{m}{M}$$

From which

$$M = \frac{m}{n}$$

$$n = 0.015 \text{ mol}$$
$$m = 0.802 \text{ g}$$
$$M = ?$$

Substituting we have

$$M = \frac{0.802\ \text{g}}{0.0150\ \text{mol}} = 53.5\ \text{g mol}^{-1}$$

Example 5

0.019 mol of the compound $Na_2CO_3.xH_2O$ weighs 5.5 g. Determine the value of x.

(H = 1, C = 12, O = 16, Na = 23)

Solution

The very first step is to calculate the molar mass of the compound as follows.

$$n = \frac{m}{M}$$

From which

$$M = \frac{m}{n}$$

n = 0.019 mol
m = 5.5 g
M = ?

Substituting we have

$$M = \frac{5.01\ \text{g}}{0.0175\ \text{mol}} = 286\ \text{g mol}^{-1}$$

Since the relative formula mass of $Na_2CO_3.xH_2O$ is 286, it then follows that

$$Na_2CO_3.xH_2O = 286$$

$$(23 \times 2) + (12 \times 1) + (16 \times 3) + x\{(1 \times 2) + (16 \times 1)\} = 286$$

So $\quad 106 + 18x = 286$

$$18x = 180$$

$$x = \frac{180}{18} = 10$$

Thus, the formula of the compound is $Na_2CO_3.10H_2O$.

Example 6

0.0411 mol of a compound XCl weighs 1.50 g. Identify the compound by matching up X with one of those given in the list of relative atomic masses below.

(H = 1, Na = 23, Cl =35.5, Ag = 108)

Solution

We know that

$$n = \frac{m}{M}$$

From which

$$M = \frac{m}{n}$$

m = 1.50 g
M = ?

Substituting we have

$$M = \frac{1.50\ \text{g}}{0.0411\ \text{mol}} = 286\ \text{g mol}^{-1}$$

Since the relative formula mass of XCl is 36.5, then

X + 35.5 = 36.5
X = 36.5 – 35.5 = 1

This is the relative atomic mass of hydrogen. Thus, the compound is HCl.

Practice Problems

1. Determine the number of moles in each of the following.
   (a) 1.0 g of calcium carbide, CaC_2.
   (b) 0.12 g of silver nitrate, $AgNO_3$.
   (c) 1.15 g of calcium carbonate, $CaCO_3$.
2. Determine the mass of each of the following.
   (a) 0.15 mol of hydrochloric acid, HCl.
   (b) 0.014 mol of zinc oxide, ZnO.
   (c) 2.5 mol of ammonium chloride, NH_4Cl.
3. Determine the molar mass of each of the following.
   (a) A substance containing 0.014 mol in 1.04 g.
   (b) A substance containing 0.11 mol in 0.20 g.
   (c) A substance containing 2.50 mol in 91.3 g.
4. 0.0121 mol of the compound Na2CO3.xH2O weighs 1.5 g. Determine the full formula of the compound.

(H = 1, C = 12, N = 14, O = 16, Na = 23, Cl = 35.5, Ca = 40, Zn = 65, Ag = 108)

CHAPTER 5: THE MOLE AND NUMBER OF PARTICLES

A mole of any substance contains approximately 6.02×10^{23} particles or elemental entities. This value is known as Avogadro's number or constant, N_A. The relationship between the number of moles in a sample of a substance and number of particles N it contains is given by

$$n = \frac{N}{N_A}$$

Example 1

Determine the number of moles in a sample of chlorine gas containing 5.98×10^{22} atoms.

$$(N_A = 6.02 \times 10^{23}\ \text{mol}^{-1})$$

Solution

We have to apply the relation

$$n = \frac{N}{N_A}$$

$N = 5.98 \times 10^{22}$

$N_A = 6.02 \times 10^{23}\ \text{mol}^{-1}$

$n = ?$

Substituting we have

$$n = \frac{5.98 \times 10^{22} \times 1\ \text{mol}}{6.02 \times 10^{23}} = 0.993\ \text{mol}$$

Example 2

A sample of hydrogen chloride contains 3.5×10^{23} molecules. Determine the number of moles in the sample.

$(N_A = 6.02 \times 10^{23}\ mol^{-1})$

Solution

We have to apply the relation

$$n = \frac{N}{N_A}$$

$$N = 3.5 \times 10^{23}$$
$$N_A = 6.02 \times 10^{23}\ mol^{-1}$$
$$n = ?$$

Substituting we have

$$n = \frac{3.5 \times 10^{23} \times 1\ mol}{6.02 \times 10^{23}} = 0.58\ mol$$

Example 3

A solution contains 0.058 mol of hydrogen ions. How many hydrogen ions are there in the solution?

$(N_A = 6.02 \times 10^{23}\ mol^{-1})$

Solution

We know that

$$n = \frac{N}{N_A}$$

$$N_A = n \times N_A$$
$$N = 6.02 \times 10^{23}\ mol^{-1}$$
$$n = 0.058\ mol^{-1}$$
$$N = ?$$

Substituting we have

$$N = 0.058 \text{ mol} \times \frac{6.02 \times 10^{23}}{1 \text{ mol}} = 3.5 \times 10^{22}$$

Example 4

How many atoms are there in 0.15 mol of oxygen, O_2?

($N_A = 6.02 \times 10^{23}$ mol^{-1})

Solution

We know that

$$n = \frac{N}{N_A}$$

$N = n \times N_A$

$N_A = 6.02 \times 10^{23}$ mol^{-1}

$n = 0.15$ mol

$N = ?$

Substituting we have

$$N = 0.15 \text{ mol} \times \frac{6.02 \times 10^{23}}{1 \text{ mol}} = 9.0 \times 10^{22}$$

Example 5

A sample of glucose, $C_6H_{12}O_6$, contains 6.05×10^{21} molecules. Determine:

(a) the number of moles in the sample;

(b) the mass of the sample.

(H = 1, C = 12, O = 16, $N_A = 6.02 \times 10^{23}$ mol^{-1})

Solution

(a) We have to apply the relation

$$n = \frac{N}{N_A}$$

$$N = 6.05 \times 10^{21}$$
$$N_A = 6.02 \times 10^{23} \text{ mol}^{-1}$$
$$n = ?$$

Substituting we have

$$n = \frac{6.05 \times 10^{21} \times 1 \text{ mol}}{6.02 \times 10^{23}} = 0.01 \text{ mol}$$

(b) We know that

$$n = \frac{m}{M}$$

$$\therefore \quad m = n \times M$$
$$M = \{(12 \times 6) + (1 \times 12) + (16 \times 6)\} \text{ g mol}^{-1}$$
$$= 180 \text{ g mol}^{-1}$$
$$n = 0.01 \text{ mol}$$
$$m = ?$$

Finally, we now substitute to obtain

$$m = 0.01 \text{ mol} \times \frac{180 \text{ g}}{1 \text{ mol}} = 1.8 \text{ g}$$

Example 6

Determine the number of molecules in 5.0 g of calcium carbonate, $CaCO_3$.

(C = 12, O = 16, Ca = 40, $N_A = 6.02 \times 10^{23} \text{ mol}^{-1}$)

Solution

The very first is to obtain the number of moles as follows.

$$n = \frac{m}{M}$$

$$M = \{(40 \times 1) + (12 \times 1) + (16 \times 3)\} \text{ g mol}^{-1}$$
$$= 100 \text{ g mol}^{-1}$$

$m = 5.0\text{ g}$

$n = ?$

Substituting we have

$$n = \frac{5.0\text{ g} \times 1\text{ mol}}{100\text{ g}} = 0.05\text{ mol}$$

Finally, we can now determine the number of molecules as follows.

$$n = \frac{N}{N_A}$$

$N = n \times N_A$

$N_A = 6.02 \times 10^{23}\text{ mol}^{-1}$

$n = 0.05\text{ mol}$

$N = ?$

Substituting we have

$$N = 0.05\text{ mol} \times \frac{6.02 \times 10^{23}}{1\text{ mol}} = 3.01 \times 10^{22}$$

Example 7

0.51 g of a metal contains 1.33×10^{22} atoms. Determine:
(a) the number of moles in the sample;
(b) the molar mass of the metal.

($N_A = 6.02 \times 10^{23}\text{ mol}^{-1}$)

Solution

(a) We have to apply the relation

$$n = \frac{N}{N_A}$$

$N = 1.33 \times 10^{22}$

$N_A = 6.02 \times 10^{23}\text{ mol}^{-1}$

$$n = ?$$

Substituting we have

$$n = \frac{1.33 \times 10^{22}\ \text{mol}}{6.02 \times 10^{23}} = 0.0221\ \text{mol}$$

(b) The molar mass is calculated as follows.

$$n = \frac{m}{M}$$

From which

$$M = \frac{m}{n}$$

$$n = 0.0221\ \text{mol}$$
$$M = ?$$

Substituting we have

$$M = \frac{0.51\ \text{g}}{0.0221\ \text{mol}} = 23\ \text{g mol}^{-1}$$

Practice Problems

1. Determine the number of moles in the following.
   (a) A sample of lead containing 4.55×10^{24} atoms.
   (b) A sample of oxygen gas containing 3.00×10^{20} atoms.
   (c) A sample of methane containing 9.08×10^{23} molecules.
2. Determine the number of particles in each of the following.
   (a) 2.5 mol of uranium.
   (b) 0.11 mol of calcium carbide.
   (c) 5.0 mol of calcium ions.
3. Determine the mass in each of the following:
   (a) 5.50×10^{23} atoms of helium gas, He.
   (b) 1.06×10^{23} molecules of carbon monoxide, CO.
   (c) 8.01×10^{23} molecules of hydrogen sulphide, H_2S.

4. Determine the molar mass of the following.
   (a) A gas containing 1.35×10^{24} atoms in 9.0 g.
   (b) A metal containing 6.02×10^{22} atoms in 6.5 g.
   (c) A gas containing 1.51×10^{23} molecules in 4.0 g.
   (H = 1, He = 4, C = 12, O = 16, S = 32, $N_A = 6.02 \times 10^{23}$ mol^{-1})

CHAPTER 6: THE MOLE AND CONCENTRATION

The relationship between the number of moles of a substance in solution and the concentration of the solution is given by

$$n = C \times V$$

where

C = concentration of solution in mole per cubic decimetre ($mol\ dm^{-3}$)

V = volume of solution in cubic decimetre (dm^3)

Note: $1000\ cm^3 = 1000\ mL = 1\ dm^3 = 1\ L$

Example 1

A beaker contains $50.00\ cm^3$ of $0.15\ mol\ dm^{-3}$ solution of sodium hydroxide. Determine the number of moles of sodium hydroxide in the beaker.

Solution

We have to apply the relation

$$n = C \times V$$
$$C = 0.15\ mol\ dm^{-3}$$
$$V = 50.00\ cm^3 = 0.050\ dm^3$$
$$n = ?$$

Substituting we have

$$n = \frac{0.15\ mol}{1\ dm^3} \times 0.050\ dm^3 = 0.0075\ mol$$

Example 2

A solution of sodium carbonate has a concentration of 0.50 mol dm^{-3}. Determine the volume of the solution that will contain 0.25 mol of the compound.

Solution

As usual, we will apply the relation

$$n = C \times V$$

From which

$$V = \frac{n}{C}$$

n = 0.25 mol
C = 0.15 mol dm^{-3}
V = ?

Substituting we have

$$V = \frac{0.25\ \text{mol} \times 1\ \text{dm}^3}{0.15\ \text{mol}} = 1.6\ \text{dm}^3$$

Example 3

50.00 cm^3 of a hydrochloric acid solution contains 0.010 mol of the acid. Determine concentration of the solution.

Solution

$n = C \times V$
n = 0.010 mol
V = 50.00 cm^3 = 0.050 dm^3
C = ?

Substituting we have

$$C = \frac{0.010\ \text{mol}}{0.050\ \text{dm}^3} = 0.20\ \text{mol dm}^3$$

Example 4

A conical flask contains 25.00 cm^3 of 0.15-mol dm^{-3} sodium hydroxide solution. Calculate:

(a) the number of moles of sodium hydroxide in the compound;

(b) the mass of sodium hydroxide in the solution.

(Na = 23, Cl = 35.5)

Solution

(a) We will apply the relation

$n = C \times V$

$C = 0.15\ \text{mol dm}^{-3}$

$V = 25\ \text{cm}^3 = 0.025\ \text{dm}^3$

$n = ?$

Substituting we have

$$n = \frac{0.15\ \text{mol}}{1\ \text{dm}^3} \times 0.025\ \text{dm}^3 = 0.0038\ \text{mol}$$

(b) Recall that

$$n = \frac{m}{M}$$

$m = n \times M$

$M = \{(23 \times 1) + (35.5 \times 1)\}\ \text{g mol}^{-1}$

$= 58.5\ \text{g mol}^{-1}$

$n = 0.00375\ \text{mol}$

$m = ?$

Substituting we have

$$m = 0.0038 \text{ mol} \times \frac{58.5 \text{ g}}{1 \text{ mol}} = 0.22 \text{ g}$$

Example 5

Determine the number of hydrogen ions in 250 cm^3 of a 0.50 mol dm^{-3} solution of the ions.

(N_A = 6.02 × 10^{23} mol^{-1})

Solution

The first step is to calculate the number of moles of hydrogen ions as follows.

$n = C \times V$
C = 0.50 mol dm^{-3}
V = 250 cm^3 = 0.25 dm^3
n = ?

Substituting we have

$$n = \frac{0.50 \text{ mol}}{1 \text{ dm}^3} \times 0.25 \text{ dm}^3 = 0.125 \text{ mol}$$

Finally, we can now calculate the number of hydrogen ions as follows.

$$n = \frac{N}{N_A}$$

From which

$N = n \times N_A$
n = 0.125 mol
N_A = 6.02 × 10^{23} mol^{-1}
N = ?

Substituting we have

Example 6

25 cm^3 of a 0.60-mol dm^{-3} solution contains 0.84 g of a base. Determine:

(a) the number of moles of the base in the solution;

(b) the molar mass of the base.

Solution

(a) We will apply the relation

$$n = C \times V$$
$$C = 0.60 \text{ mol dm}^{-3}$$
$$V = 25 \text{ cm}^3 = 0.025 \text{ dm}^3$$
$$n = ?$$

Substituting we have

$$n = \frac{0.60 \text{ mol}}{1 \text{ dm}^3} \times 0.025 \text{ dm}^3 = 0.015 \text{ mol}$$

(b) Recall that

$$n = \frac{m}{M}$$

From which

$$M = \frac{m}{n}$$

$$m = 0.84 \text{ g}$$
$$n = 0.015 \text{ mol}$$
$$M = ?$$

Substituting we have

$$M = \frac{0.84 \text{ g}}{0.015 \text{ mol}} = 56 \text{ g mol}^{-1}$$

Practice Problems

1. Determine the number of moles of the following.
   (a) 25.00 cm^3 of 0.011mol dm^{-3} solution of potassium

hydroxide.

(b) 150 cm^3 of 0.25 mol dm^{-3} solution of silver nitrate.

(c) 20.00 cm^3 of 0.10 mol dm^{-3} solution of sodium hydroxide.

2. Determine the volume of the following.

(a) 2.5 mol of 0.22-mol dm^{-3} solution of sodium nitrate.

(b) 0.0050 mol of 0.15 mol dm^{-3} solution of nitric acid.

(c) 0.75 mol of 1.15 mol dm^{-3} solution of copper(II) sulfate.

3. Determine the concentration of the following.

(a) 50.00 cm^3 of a solution containing 0.050 mol of ethanoic acid.

(b) 250 cm^3 of a solution containing 1.5 mol of ethanol.

(c) 150 cm^3 of a solution containing 0.10 mol of hydrochloric acid.

4. Determine the number of particles in the following.

(a) 100.00 cm^3 of a 1.5-mol dm^{-3} solution of chloride ions.

(b) 150 cm^3 of 0.55-mol dm^{-3} solution of carbonate ions.

(c) 250 cm^3 of a 0.11-mol dm^{-3} solution of sucrose.

5. Determine the molar mass of the substance in the following.

(a) 500.00 cm^3 of 1.00-mol dm^{-3} solution containing 53 g of a substance.

(b) 150 cm^3 of 0.66 mol dm^{-3} solution containing 4.0 g of a substance.

(c) 1.5 dm^3 of 0.10-mol dm^{-3} solution containing 14.7 g of a substance.

(H = 1, N = 14, O = 16, Na = 23, K = 39, Ag = 108; $N_A = 6.02 \times 10^{23}$ mol^{-1})

CHAPTER 7: THE MOLE AND MOLAR VOLUME

The volume and number of moles of gases vary with temperature and pressure. These two quantities are thus reported at some standard reference conditions of temperature and pressure. Every gas occupies a volume of approximately 22.4 dm^3 at standard temperature and pressure, s.t.p. This volume of a gas is called molar gas volume, V_m, at s.t.p. Standard temperature and pressure correspond to 273 K (0°C) and 760 mmHg (1 atm), respectively.

The relationship between the number of moles of a gas and its volume, V, as s.t.p. is given by

$$n = \frac{V}{V_m}$$

Example 1

Calculate the number of moles in 250 cm^3 of oxygen at s.t.p.

(V_m at s.t.p. = 22.4 dm^3 mol^{-1})

Solution

We will apply the relation

$$n = \frac{V}{V_m}$$

V = 250 cm^3 = 0.25 dm^3

V_m = 22.4 dm^3 mol^{-1}

n = ?

Substituting we have

$$n = \frac{0.25\ \text{dm}^3 \times 1\ \text{mol}}{22.4\ \text{dm}^3} = 0.011\ \text{mol}$$

Example 2

Determine the volume of 0.25 mol of carbon dioxide at s.t.p.

(V_m at s.t.p. = 22.4 $dm^3\ mol^{-1}$)

Solution

We know that

$$n = \frac{V}{V_m}$$

From which

$$V = n \times V_m$$

$$n = 0.25\ \text{mol}$$

$$V_m = 22.4\ \text{dm}^3\ \text{mol}^{-1}$$

$$V = ?$$

Substituting we have

$$V = 0.25\ \text{mol} \times \frac{22.4\ \text{dm}^3}{1\ \text{mol}} = 5.6\ \text{dm}^3$$

Example 3

Determine the number of molecules in 550 cm^3 of carbon dioxide at s.t.p.

(V_m at s.t.p.= 22.4 $dm^3\ mol^{-1}$, N_A = $6.02 \times 10^{23}\ mol^{-1}$)

Solution

The very first step is to calculate the number of moles CO_2.

$$n = \frac{V}{V_m}$$

$$V = 550\ \text{cm}^3 = 0.55\ \text{dm}^3$$

$V_m = 22.4$ dm3 mol-1

$n = ?$

Substituting we have

$$n = \frac{0.55\ \text{dm}^3 \times 1\ \text{mol}}{22.4\ \text{dm}^3} = 0.02455\ \text{mol}$$

We can now calculate the number of molecules as follows.

$$n = \frac{N}{N_A}$$

$N = n \times N_A$

$n = 0.02455$ mol

$N_A = 6.02 \times 10^{23}$ mol^{-1}

$N = ?$

Substituting we have

$$N = 0.02455 \times \frac{6.02 \times 10^{23}}{1\ \text{mol}} = 1.5 \times 10^{22}$$

Example 4

Determine the mass of 150 cm^3 of hydrogen sulfide, H_2S, at s.t.p.
(H = 1, S = 32, V_m at s.t.p. = 22.4 dm^3 mol^{-1})

Solution

The very first step is to calculate the number of moles as follows.

$$n = \frac{V}{V_m}$$

$V = 150$ cm^3 = 0.15 dm^3

$V_m = 22.4$ dm^3 mol^{-1}

$n = ?$

Substituting we have

$$n = \frac{0.15\ \text{dm}^3 \times 1\ \text{mol}}{22.4\ \text{dm}^3} = 0.006696\ \text{mol}$$

The final step is to calculate the mass as follows.

$$n = \frac{m}{M}$$

From which

$$m = n \times M$$
$$M = \{(1 \times 2) + (32 \times 1)\}\ \text{g mol}^{-1}$$
$$= 34\ \text{g mol}^{-1}$$
$$n = 0.006696\ \text{mol}$$
$$m = ?$$

We will now substitute to obtain

$$m = 0.006696\ \text{mol} \times \frac{34\ \text{g}}{1\ \text{mol}} = 0.23\ \text{g}$$

Example 5

55.0 g of a gas occupies a volume of 1400 cm^3 at s.t.p. Calculate the molar mass of the gas.

(V_m at s.t.p. = 22.4 $\text{dm}^3\ \text{mol}^{-1}$)

Solution

The first step is to calculate the number of moles of the gas.

$$n = \frac{V}{V_m}$$

$$V = 1400\ \text{cm}^3 = 1.4\ \text{dm}^3$$
$$V_m = 22.4\ \text{dm}^3\ \text{mol}^{-1}$$
$$n = ?$$

Substituting we have

$$n = \frac{1.4\ \text{dm}^3 \times 1\ \text{mol}}{22.4\ \text{dm}^3} = 0.0625\ \text{mol}$$

We can now calculate the molar mass as follows.

$$n = \frac{m}{M}$$

From which

$$M = \frac{m}{n}$$

m = 5.0 g
n = 0.0625 mol
M = ?

Substituting we have

$$M = \frac{5.0\ \text{g}}{0.0625\ \text{mol}} = 80\ \text{g mol}^{-1}$$

Practice Problems

1. Determine the number of moles at s.t.p. in each of the following.
   (a) 550 cm^3 of neon.
   (b) 250 cm^3 of sulfur dioxide.
   (c) 1500 cm^3 of methane.
2. Determine the volume at s.t.p. in each of the following:
   (a) 2.5 mol of hydrogen.
   (b) 0.15 mol of nitrogen.
   (c) 1.25 mol of benzene.
3. Determine the number of molecules at s.t.p. in each of the following.
   (a) 5500 cm^3 of nitrogen dioxide.
   (b) 150 cm^3 of carbon monoxide.
   (c) 250 cm^3 of hydrogen sulfide.
4. Determine the mass at s.t.p. of each of the following.

(a) 250 cm^3 of sulfur trioxide, SO_3.
(b) 2500 cm3 of nitrogen, N_2.
(c) 510 cm^3 of hydrogen chloride, HCl.

5. Determine the molar mass of the gas in each of the following.
(a) 7.0 g occupying a volume of 5600 cm^3 at s.t.p.
(b) 4.5 g occupying a volume of 2960 cm^3 at s.t.p.
(c) 2.0 g occupying a volume of 2800 cm^3 at s.t.p.
(H = 1, N = 14, O = 16, S = 32, Cl =35.5, N_A = 6.02 × 10^{23} mol^{-1}, V_m at s.t.p. = 22.4 dm^3 mol^{-1})

CHAPTER 8: EMPIRICAL AND MOLECULAR FORMULAE

An empirical formula is the simplest formula of a substance that shows its constituent elements and the whole-number ratio in which their atoms are combined. On the other hand, a molecular formula is the unique formula of a compound which shows its constituent elements and the actual number of atoms of each element in its formula unit.

The molecular formulae of benzene and ethyne, for example, are C_6H_6 and C_2H_2 respectively. However, both compounds have the same empirical formula of CH.

The molecular formula of a compound is equivalent to its formula unit or chemical formula. Note that molecular formulae are used in relation to only molecular compounds. Ionic compounds are usually denoted by formulae involving the simplest ratios of atoms. Thus, the empirical formulae of ionic compounds are effectively their formula units or chemical formulae.

Empirical formulae are determined from mass or percent compositions of compounds as follows.

1. **Calculate the number of moles**
Determine the number of moles of atoms of each element in the compound.

2. **Determine the smallest whole-number mole ratio**
The simplest whole-number mole ratio of atoms should be

determined by dividing each number of moles by the least value. Values should be rounded to the nearest whole number provided the difference is not greater than 5%. This implies that numbers with decimals from 0.9 and above should be rounded up to the nearest whole number, while those with 0.1 and below should be rounded down to the nearest whole number. Other values should be converted to fraction and multiplied by their lowest common multiple (LCM) to obtain the whole-number mole ratio of atoms.

Some examples of decimals often encountered in empirical formula calculations and their equivalent fractions, are as follows.

$$0.67 = \frac{2}{3} \text{ : e.g. } 2.67 = 2 + \frac{2}{3} = \frac{8}{3}$$

$$0.25 = \frac{1}{4} \text{ : e.g. } 2.25 = 2 + \frac{1}{4} = \frac{9}{3}$$

$$0.50 = \frac{1}{2} \text{ : e.g. } 2.50 = 2 + \frac{1}{2} = \frac{5}{2}$$

3. Write the empirical formula

Write out the empirical formula of the compound by using the numbers obtained in Step 2 as subscripts of the appropriate atoms.

Determination of molecular formula from empirical formula

Once the empirical formula of a compound is known, its molecular formula is obtained by calculating the number of empirical formula units in the molecular formula as follows.

$$\text{(Empirical formula mass)}x = M_r$$

where x is the number of empirical formula units in a molecular formula.

The value obtained is multiplied by the subscript of each atom in the empirical formula.

Example 1

Determine empirical formula of ethene, C_2H_4.

Solution

The molecular formula of the compound is C_2H_4. The subscripts of the atoms in the molecular formula have a common factor of 2; hence, we have to divide through by this number to obtain CH_2, which is the required empirical formula.

Example 2

Determine the empirical formula of sodium carbonate if its chemical formula is Na_2CO_3.

Solution

The subscripts have no common factor, i.e., the number of atoms of each element is already in its simplest form. In this case the chemical and empirical formulae are the same. Thus, the required empirical formula is Na_2CO_3. As noted earlier, this is the case with all ionic compounds.

Example 3

Determine the empirical formula of ethanol if its molecular formula is C_2H_5OH.

Solution

For clarity, we can also write the molecular formula as C_2H_6O. It is obvious that the empirical formula is also C_2H_5OH or C_2H_6O.

Example 4

On analysis, a sample of a hydrocarbon yields 18.5 g of carbon and 1.5 g of hydrogen. Determine the empirical formula of the compound.

(H = 1, C = 12)

Solution

The first step is to calculate the number of moles of each atom as follows.

$$n = \frac{m}{M}$$

For carbon we have

m = 18.5 g
M = 12 g mol^{-1}
n = ?

Substituting we have

$$n = \frac{18.5\ g \times 1\ mol}{12\ g} = 1.5\ mol$$

For hydrogen we have

m = 1.5 g
M = 1 g mol^{-1}
n =?

Substituting we have

$$n = \frac{1.5\ g \times 1\ mol}{1\ g} = 1.5\ mol$$

The next step is to determine the whole number ratio of atoms by dividing each value by1.5 mol.

$$C = \frac{1.5\ \text{mol}}{1.5\ \text{mol}} = 1$$

$$H = \frac{1.5\ \text{mol}}{1.5\ \text{mol}} = 1$$

Thus, the empirical formula of the compound is as CH.

Example 5

A hydrated salt has the following composition: sodium (Na), 16%; carbon (C), 4.20%; oxygen (O), 16.8%; water of crystallisation (H_2O), 62.9%. Determine the empirical formula of the compound.

(H = 1, C = 12, O = 16, Na = 23)

Solution

The percent composition can be expressed in grams by assuming a 100-g sample of the compound. A usual, we will begin by calculating the number of moles of each atom as follows.

$$n = \frac{m}{M}$$

For sodium atoms we have

$m = 16.1\ \text{g}$
$M = 23\ \text{g mol}^{-1}$
$n = ?$

Substituting we have

$$n = \frac{16.1\ \text{g} \times 1\ \text{mol}}{23\ \text{g}} = 0.70\ \text{mol}$$

For carbon atoms we have

$m = 4.20\ \text{g}$
$M = 12\ \text{g mol}^{-1}$
$n = ?$

Substituting we have

$$n = \frac{4.20\ \text{g} \times 1\ \text{mol}}{12\ \text{g}} = 0.35\ \text{mol}$$

For oxygen atoms we have

m = 16.8 g
M = 16 g mol^{-1}
n = ?

Substituting we have

$$n = \frac{16.8\ \text{g} \times 1\ \text{mol}}{16\ \text{g}} = 1.05\ \text{mol}$$

For water of crystallisation, H_2O, we have

m = 62.9 g
M = {(1× 2) + (16 × 1)} g mol^{-1}
= 18 g mol^{-1}
n = ?

Substituting we have

$$n = \frac{62.9\ \text{g} \times 1\ \text{mol}}{18\ \text{g}} = 3.49\ \text{mol}$$

The next step is to divide each value by 0.35 mol as follows.

$$\text{Na} = \frac{0.70\ \text{mol}}{0.35\ \text{mol}} = 2$$

$$\text{C} = \frac{0.35\ \text{mol}}{0.35\ \text{mol}} = 1$$

$$\text{O} = \frac{1.05\ \text{mol}}{0.35\ \text{mol}} = 3$$

$$\text{H}_2\text{O} = \frac{3.49\ \text{mol}}{0.35\ \text{mol}} = 10$$

Thus, the empirical formula of the compound is $Na_2CO_3.10H_2O$.

Example 6

The percentage by mass of hydrogen in a hydrocarbon is 11.1%. Determine the empirical formula of the compound.

(H = 1, C = 12)

Solution

As usual, we will assume a 100-g sample of the compound. As usual, we will begin with calculating the number of moles of each atom.

$$n = \frac{m}{M}$$

For hydrogen atoms we have

m = 11.1 g
M = 1 g mol^{-1}
n = ?

Substituting we have

$$H_2O = \frac{3.49\ \text{mol}}{0.35\ \text{mol}} = 10$$

For carbon atoms we have

m = 100 g – 11.1 g = 88.9 g
M = 12 g mol^{-1}
n = ?

Substituting we have

$$n = \frac{88.9\ \text{g} \times 1\ \text{mol}}{12\ \text{g}} = 7.4\ \text{mol}$$

The next step is to divide the values by 7.4 mol.

$$C = \frac{7.4 \text{ mol}}{7.4 \text{ mol}} = 1$$

$$H = \frac{11.1 \text{ mol}}{7.4 \text{ mol}} = 1.5 = \frac{3}{2}$$

The simplest whole-number mole ratio of atoms is obtained by multiplying the two values by 2, i.e., the LCM of 1 and 2, to obtain

C = 2

H = 3

Thus, the empirical formula of the compound is C_2H_3.

Example 7

The empirical formula of a compound is CH. Determine its molecular formula if its relative molecular mass is 78.

(H = 1, C = 12)

Solution

We will apply the relation

(Empirical formula mass)$x = M_r$

Empirical formula = CH

$M_r = 78$

$x = ?$

Substituting we have

(Mass of CH)$x = 78$

$(12 + 1)x = 78$

$13x = 78$

$$x = \frac{78}{13} = 6$$

Finally, we must now multiply the number of atoms of in the empirical formula by 6 to obtain the molecular formula C_6H_6.

Example 8

The empirical formula of a compound is CH_2O. Determine its molecular formula if its molar mass is 180 g mol^{-1}.

(H = 1, C = 12, O = 16)

Solution

$$(\text{Empirical formula mass})x = M_r$$
$$\text{Empirical formula} = CH_2O$$
$$M_r = 180$$
$$x = ?$$

Substituting we have

$$(\text{Mass of } CH_2O)x = 180$$
$$\{(12 \times 1) + (1 \times 2) + (16 \times 1)\}x = 180$$

So

$$30x = 180$$

$$x = \frac{180}{30} = 6$$

Thus, the molecular formula of the compound is $C_6H_{12}O_6$.

Example 9

The empirical formula of a hydrocarbon is C_7H_8. Determine its molecular formula if 0.25 mol of the compound weighs 23 g.

(H = 1, C = 12, O = 16)

Solution

As usual, we will apply the relation

$$(\text{Empirical formula mass})x = M_r$$
$$\text{Empirical formula} = C_7H_8$$
$$M_r = ?$$
$$x = ?$$

The relative molecular mass is obtained as follows.

$$n = \frac{m}{M}$$

From which

$$M = \frac{m}{n}$$

m = 23 g

n = 0.25 mol

M = ?

Substituting we have

$$M = \frac{23\ \text{g}}{0.25\ \text{mol}} = 92\ \text{g mol}^{-1}$$

$$M_r = 92$$

Substituting we have

$$(\text{Mass of } C_7H_8)x = 92$$

$$\{(12 \times 7) + (1 \times 8)\}x = 92$$

$$92x = 92$$

$$x = \frac{92}{92} = 1$$

There is only one empirical formula unit in the molecular formula; hence, the empirical and molecula formulae of the compound are both C_7H_8.

Practice Problems

1. Determine the empirical formulae of the compounds with the following molecular formulae.
   (a) C_5H
   (b) C_4H_{10}
   (c) $CaCO_3$
   (d) C_2H_5COOH
   (e) $C_8H_{10}N_4O_2$
2. Determine the empirical formulae of the compound with the following percent compositions.
   (a) C = 80%, H = 20%
   (b) C = 27.3%, O = 72.7%
   (c) C = 52.2%, H = 13.0%, O = 34.8%
   (d) Cu = 25.4%, S = 12.8%, O = 25.7%, H_2O = 36.1%

(e) C = 73%, H = 5.4%, O = 21.6%

3. Determine the molecular formulae of the compounds whose empirical formulae and relative molecular masses are given below.
(a) $C_{10}H_{22}$, 142
(b) CHO_2, 90
(c) H_2CO_2, 46
(d) C_3H_6O, 58
(e) CH_2O, 180

(H = 1, C =12, O = 16, S = 32, Cu = 63.5)

4. 0.50 mol of a mono-alkanoic acid weighs 44 g. Determine the molecular formula and name of the acid.

(H = 1, C = 12, O = 16)

5. The analysis of a hydrocarbon sample contains 87% carbon. Given that 0.25 mol of the compound weighs 69 g, determine:
(a) the molar mass of the compound;
(b) the empirical formula of the compound;
(c) the molecular formula of the compound.

(H = 1, C = 12)

6. The combustion of a 1.5-g sample of lycopene, the compound responsible for the colour of ripe tomatoes, yields 4.93 g of carbon dioxide. Given that 0.0225mol of the compound weighs 12.06 g, determine:
(a) its molar mass;
(b) its mass percentage composition;
(c) its empirical formula;
(d) its molecular formula.

(H = 1, C = 12)

CHAPTER 9: CHEMICAL EQUATIONS

A chemical equation is a short way of writing a chemical reaction using only chemical symbols and formulae of reactants and products. In a chemical equation, an arrow points from the reactants to the products. A reversible reaction is indicated with double-headed arrows ($\rightleftharpoons$).

Some examples of chemical equations are given below.

1. The equation for the decomposition of calcium carbonate into calcium oxide and carbon dioxide.
   $$CaCO_3(s) \rightarrow CaO(s) + CO_2(g)$$
2. The equation for the reaction between hydrogen and oxygen gases to produce steam.
   $$2H_2(g) + O2(g) \rightarrow 2H_2O(g)$$
3. The reversible reaction between nitrogen and hydrogen to produce ammonia.
   $$N_2(g) + 3H_2(g) \rightleftharpoons 2NH_3(g)$$

Balancing chemical equations

Mass is conserved in ordinary chemical reactions. This implies that the numbers of each type of atom on both sides of a chemical equation must be equal. A chemical equation that is written in this way is called a balanced equation. Consider, for example, the following equation.

$$2H_2(g) + O_2(g) \rightarrow 2H_2O(g)$$

There are 4 hydrogen atoms and 2 oxygen atoms on the

reactant side, while there are equal numbers on the product side. Thus, the equation is balanced.

An unbalanced equation must be balanced by placing appropriate numbers, called coefficients, to the left of relevant reactants and products.

Example 1

Balance the following equation.

$$2CO(g) + O_2(g) \rightarrow 2CO_2(g)$$

(c) $CH_4(g) + Cl_2(g) \rightarrow CH_3Cl(g) + H_2(g)$
(d) $2HCl(aq) + Zn(s) \rightarrow ZnCl_2(aq) + H_2(g)$
(e) $C_6H_{12}O_6(s) + O_2(g) \rightarrow CO_2(g) + 6H_2O(l)$

Solution

We have to be sure that the number of each type of atom on the left side of the arrow matches the number on the right side.

$$2CO(g) + O_2(g) \rightarrow 2CO_2(g)$$

The numerical value of an atom in a substance appearing in a chemical equation is obtained by multiplying the coefficient of the substance by the subscript of the atom; hence, in this equation, there are 2 and 4 carbon and oxygen atoms on both sides of the equation. The atom count is as follows.

	LHS	RHS
C	$2 \times 1 = 2$	$2 \times 1 = 2$
O	$(2 \times 1) + (1 \times 2) = 4$	$2 \times 2 = 4$

Thus, the equation is balanced as written.

Example 2

Balance the following equation.

$$H_2(g) + I_2(g) \rightarrow HI(g)$$

Solution

The atom count is as follows.

	LHS	RHS
H	$1 \times 2 = 2$	$1 \times 1 = 1$
I	$1 \times 2 = 2$	$1 \times 1 = 1$

Thus, the equation is unbalanced. The equation is balanced by adding a 2 to the left of HI, i.e.

$$H_2(g) + I_2(g) \rightarrow 2HI(g).$$

Example 3

Balance the following equation.

$$CH_4(g) + Cl_2(g) \rightarrow CH_3Cl(g) + H_2(g)$$

Solution

The atom count is as follows.

	LHS	RHS
C	$1 \times 1 = 1$	$1 \times 1 = 1$
H	$1 \times 4 = 4$	$(1 \times 3) + (1 \times 2) = 5$
Cl	$1 \times 2 = 2$	$2 \times 1 = 2$

The equation is obviously not balanced. To balance the equation we have to add a 2 to the left of CH_4 and CH_3Cl to obtain

$$2CH_4(g) + Cl_2(g) \rightarrow 2CH_3Cl(g) + H_2(g)$$

Example 4

Balance the following equation.

$$2HCl(aq) + Zn(s) \rightarrow ZnCl_2(aq) + H_2(g)$$

Solution

The atom count is as follows.

	LHS	RHS
H	$2 \times 1 = 2$	$1 \times 2 = 2$

Cl	$2 \times 1 = 2$	$1 \times 2 = 2$
Zn	$1 \times 1 = 1$	$1 \times 1 = 1$

Thus, the equation is already balanced as written.

Example 5

$$C_6H_{12}O_6(s) + O_2(g) \rightarrow CO_2(g) + 6H_2O(l)$$

The atom count is as follows.

	LHS	RHS
C	$1 \times 6 = 6$	$1 \times 1 = 1$
H	$1 \times 12 = 12$	$6 \times 2 = 12$
O	$(1 \times 6) + (1 \times 2) = 8$	$(1 \times 2) + (6 \times 1) = 8$

The equation is definitely unbalanced. The equation will be balanced if we add a 6 to the left of both O_2 and CO_2. Thus, the balanced equation is

$$C_6H_{12}O_6(s) + 6O_2(g) \rightarrow 6CO_2(g) + 6H_2O(l)$$

Information from chemical equations

A balanced chemical equation provides the following information.

1. The reactants and products through their symbols and formulae.
2. The direction of a reaction.
3. The physical states of substances are denoted in parentheses next to their symbols or formulae. A gas is denoted ‘g’, a liquid, ‘l’, a solid, ‘s’, and an aqueous solution of a substance, ‘aq’.
4. The mole ratio of reactants to products. The number of moles of each substance in a chemical equation is given by its stoichiometric (numerical) coefficient in the equation, e.g.

$$2H_2(g) + O_2(g) \rightarrow 2H_2O(g)$$

	$2H_2(g)$	$O_2(g)$	$2H_2O(g)$
Moles	2	1	2
Mole ratio	2 :	1 :	2

4. Sometimes, the conditions required for the reaction to occur or proceed faster, such as heat change or catalyst, can be stated, e.g.

$$2SO_2(g) + O_2(g) \xrightarrow{V_2O_5} 2SO_3(g) \qquad \Delta H = -198.2\ kJ$$

The reaction shows that the conversion of sulfur dioxide to sulfur trioxide is catalysed by vanadium(V) oxide and that the reaction gives off 99.1 kJ of heat per mole of sulfur trioxide produced.

Practice Problems

Check each of the following equations carefully and balance those that are unbalanced.

(1) $2H_2O_2(aq) \rightarrow H_2O(l) + O_2(g)$

(2) $CH_4(g) + O_2(g) \rightarrow CO_2(g) + H_2O(g)$

(3) $CaCO_3(s) + 2HCl(aq) \rightarrow CaCl_2(aq) + CO_2(g) + H_2O(l)$

(4) $Na_2CO_3(aq) + H_2SO_4(aq) \rightarrow Na_2SO_4(aq) + CO_2(g) + H_2O(l)$

(5) $2Na(s) + 2H_2O(l) \rightarrow NaOH(aq) + H_2(g)$

(6) $C_5H_{12}(g) + O2(g) \rightarrow CO_2(g) + H_2O(g)$

(7) $Ca(OH)_2(aq) + H_3PO_4(aq) \rightarrow Ca_3(PO_4)_2(aq) + H_2O(l)$

(8) $2AgI(aq) + Na_2S(s) \rightarrow Ag_2S(s) + NaI(aq)$

(9) $3CaCl_3(aq) + 2Na_3PO_4(aq) \rightarrow Ca_3(PO_4)_2(aq) + 6NaCl(aq)$

(10) $4FeS(s) + 7O_2(g) \rightarrow 2Fe_2O_3(s) + 4SO_2(g)$

CHAPTER 10: CALCULATIONS BASED ON CHEMICAL EQUATIONS

We have earlier seen that the mole ratio of reactants to products is one of the information obtainable from a balanced chemical equation. In fact, one of the characteristics of a chemical reaction is that it occurs between fixed amounts of reactants to produce fixed amounts of products. This quantitative relationship that exists between the amounts of reactants and products in a balanced chemical equation is called stoichiometry. The knowledge of stoichiometry is used for calculating the amount of a desired substance that would be produced from some given amounts of reactants, and vice versa.

Example 1

What mass of oxygen would be produced from the complete decomposition of 5.0 mol of hydrogen peroxide? The equation of reaction is as follows.

$$2H_2O_2(aq) \rightarrow 2H_2O(l) + O_2(g)$$

$$(H = 1, O = 16)$$

Solution

Let the amount of O_2 that would be produced from 5.0 mol of H_2O_2 be n. According to the equation, 2 mol of H_2O_2 would produce 1 mol of O2; hence,

$$n = 5.0 \text{ mol}$$

$$1 \text{ mol} = 2 \text{ mol}$$

So $$n \times 2 \text{ mol} = 1 \text{ mol} \times 5 \text{ mol}$$

$$n = \frac{1\ \text{mol} \times 5\ \text{mol}}{2\ \text{mol}} = 2.5\ \text{mol}$$

∴ $m = n \times M$

$n = 2.5$ mol

$M = (16 \times 2)$ g mol−1 = 32 g mol^{-1}

$m = ?$

We will now substitute to obtain

$$m = 2.5\ \text{mol} \times \frac{32\ \text{g}}{1\ \text{mol}} = 80\ \text{g}$$

Example 2

On heating, calcium carbonate decomposes as follows.

$CaCO_3(s) \rightarrow CaO(s) + CO_2(g)$

How many moles of calcium carbonate are required for the production of 3.5 g of calcium oxide?

(C = 12, O = 16, Ca = 40)

Solution

Let the mass of $CaCO_3$ required to produce 3.5 g of CaO be *m*. According to the equation, 100 g of $CaCO_3$ would produce 56 g of CaO; hence,

$m = 3.5$ g

100 g = 56 g

So $m \times 56\ \text{g} = 3.5\ \text{g} \times 100\ \text{g}$

$$m = \frac{3.5\ \text{g} \times 100\ \text{g}}{56\ \text{g}} = 6.25\ \text{g}$$

Finally, we will now calculate the number of moles as follows.

$$n = \frac{m}{M}$$

$M = \{40 + 12 + (3 \times 16)\}\ \text{g mol}^{-1}$

$= 100\ \text{g mol}^{-1}$

$n = ?$

We will now substitute to obtain

$$n = \frac{6.25\ \text{g} \times 1\ \text{mol}}{100\ \text{g}} = 0.063\ \text{mol}$$

Example 3

Consider the following reaction.

$$4FeS(s) + 7O2(g) \rightarrow 2Fe_2O_3(s) + 4SO_2(g)$$

Determine the amount of oxygen, in moles, required to burn 1.5 mol of iron(II) sulfide completely; hence, calculate the mass of iron(II) sulfide required to react completely with 2.5 mol of iron(II) sulfide.

(O = 16, S = 32, Fe = 56)

Solution

Let the amount of O_2 required to completely burn 1.5 moles of FeS be n. According to the equation, it requires 7 mol of O2 to burn 4 mol of FeS; hence,

$$n = 1.5\ \text{mol}$$

$$7\ \text{mol} = 4\ \text{mol}$$

$$n \times 4\ \text{mol} = 1.5\ \text{mol} \times 7\ \text{mol}$$

$$n = \frac{1.5\ \text{mol} \times 7\ \text{mol}}{4\ \text{mol}} = 2.6\ \text{mol}$$

Similarly, let the number of moles of FeS required to react with 2.5 mol of O_2 be m. According to the equation, it requires 4 mol of FeS to react with 7 mol of O2; hence,

$$n = 2.5\ \text{mol}$$

$$4\ \text{mol} = 7\ \text{mol}$$

$$n \times 7\ \text{mol} = 2.5\ \text{mol} \times 4\ \text{mol}$$

$$n = \frac{2.5\ \text{mol} \times 4\ \text{mol}}{7\ \text{mol}} = 1.43\ \text{mol}$$

Finally, we have to convert this amount of FeS to mass as follows.

$$n = \frac{m}{M}$$

$m = n \times M$
$n = 2.5$ mol
$M = (56 + 32)$ g mol^{-1} = 88 g mol^{-1}
$m = ?$

Substituting we have

$$m = 1.43 \text{ mol} \times \frac{88 \text{ g}}{1 \text{ mol}} = 126 \text{ g}$$

Example 4

Methane burns in air as follows.

$CH_4(g) + 2O_2(g) \rightarrow CO_2(g) + 2H_2O(l)$

What volume of carbon dioxide would be produced from the complete burning of 1.5 kg of methane at s.t.p.?

(H = 1, C = 12, O = 16, V_m at s.t.p. = 22.4 dm^3 mol^{-1})

Solution

We wil begin by calculating the number of moles of CO_2 produced at s.t.p. Let the amount of CO_2 produced from 93.75 mol of CH_4 be n. According to the equation, 1 mol of CH_4 would produce 1 mol of CO_2; hence,

$n = 93.75$ mol
1 mol = 1 mol
$n \times 1 = 93.75$ mol $\times$ 1 mol

$$n = \frac{1500 \text{ g} \times 1 \text{ mol}}{16 \text{ g}} = 93.75 \text{ mol}$$

Finally, we can now calculate the volume at s.t.p. as follows.

$$n = \frac{V}{22.4 \text{ dm}^3 \text{ mol}^{-1}}$$

From which,

$$V = n \times 22.4 \text{ dm}^3 \text{ mol}^{-1}$$

n = 2100 dm

V = ?

Substituting we have

$$V = 93.75 \text{ mol} \times \frac{22.4 \text{ dm}^3}{1 \text{ mol}} = 2100 \text{ dm}^3$$

Example 5

Determine the mass of sodium sulfate, Na_2SO_4, produced when 50.0 cm^3 of 0.10 mol dm^{-3} solution sulfuric acid, H_2SO_4, is mixed with excess sodium hydroxide, NaOH, solution. The equation of reaction is as follows.

$$H_2SO_4(aq) + 2NaOH(aq) \rightarrow Na_2SO_4(aq) + 2H_2O(l)$$

(H = 1, O = 16, Na = 23, S = 32)

Solution

Let the mass of Na_2SO_4 produced from 0.49 g of H_2SO_4 be m. According to the equation, 98 g of H_2SO_4 would produce 142 g of Na_2SO_4; hence,

m = 0.49 g

142 g = 98 g

$m \times 98$ g = 0.49 g × 142 g

$$m = \frac{0.49 \text{ g} \times 142 \text{ g}}{98 \text{ g}} = 0.71 \text{ g}$$

Example 6

Determine the volume of a 0.25-mol dm^{-3} hydrochloric acid solution required for the complete reaction of the calcium oxide produced from the complete decomposition of 4.5 g of calcium carbonate. The equations of reaction are as follows.

$$CaCO_3(s) \rightarrow CaO(s) + CO_2(g)$$

$$CaO(s) + 2HCl(aq) \rightarrow CaCl_2(aq) + H_2O(l)$$

(C = 12, O = 16, Ca = 40)

Solution

We will begin by calculating the amount of CaO produced as

follows. Let the mass of CaO produced from 4.5 g of $CaCO_3$ be m. According to the equation, 100 g of $CaCO_3$ would produce 56 g of CaO; hence,

$$m = 4.5\text{ g}$$
$$56\text{ g} = 100\text{ g}$$
$$m \times 100\text{ g} = 4.5\text{ g} \times 56\text{ g}$$
$$m = \frac{4.5\text{ g} \times 56\text{ g}}{100\text{ g}} = 2.52\text{ g}$$

We will now convert 2.52 g of CaO to moles.

$$n = \frac{m}{M}$$

$$m = 2.52\text{ g}$$
$$M = \{(40 \times 1) + (16 \times 1)\}\text{ g mol}^{-1}$$
$$= 56\text{ g mol}^{-1}$$
$$n = ?$$

Substituting we have

$$n = \frac{2.52\text{ g} \times 1\text{ mol}}{56\text{ g}} = 0.045\text{ mol}$$

The next step is to determine the number of moles of HCl required for the complete neutralisation of 0.045 mol of CaO. Let the amount of HCl required to neutralise 0.045 mol of CaO be n. According to the equation, it requires 2 mol of HCl to neutralise 1 mol of CaO; hence,

$$n = 0.045\text{ mol}$$
$$2\text{ mol} = 1\text{ mol}$$
$$n \times 1\text{ mol} = 0.045\text{ mol} \times 2\text{ mol}$$
$$n = \frac{0.045\text{ mol} \times 2\text{ mol}}{1\text{ mol}} = 0.090\text{ mol}$$

Finally, we have to determine the volume of 0.090 mol of a 0.25-mol dm^{-3} solution of HCl.

$$n = C \times V$$

From which

$$n = 0.090\text{ mol}$$

$C = 0.25 \text{ mol dm}^{-3}$

$V = ?$

Substituting we have

$$V = \frac{n}{C}$$

Practice Problems

1. Iron reacts with chlorine as follows.

   $2Fe(s) + 3Cl_2(g) \rightarrow 2FeCl_3(s)$

   Determine the number of moles of chlorine required for the complete reaction of 5.5 mol of iron.
2. Determine the number of moles of hydrogen require for the production of 4.0 mol of ammonia gas. The equation of reaction is as follows.

   $N_2(g) + 3H_2(g) \rightleftharpoons 2NH_3(g)$.
3. Consider the following reaction.

   $AlCl_3(s) + 3NaOH(aq) \rightarrow Al(OH)_3(s) + 3NaCl(aq)$

   Determine the mass of aluminium chloride required for the complete reaction of 1.25 mol of sodium hydroxide.

   (Al = 27, Cl = 35.5)
4. 2.5 mol of potassium permanganate, $KMnO_4$, was mixed with excess hydrochloric acid, HCl, solution. Determine the mass of potassium chloride, KCl, produced. The equation of reaction is as follows.

   $16HCl(aq) \rightarrow 2KCl(aq) + 2MnCl_2(aq) + 8H_2O(l) + 5Cl_2(g)$

   (Cl = 35.5, K = 39)
5. Sodium peroxide reacts with water at room temperature as follows.

   $2Na_2O_2(aq) + 2H_2O(l) \rightarrow 4NaOH(aq) + O_2(g)$

   Determine the volume of oxygen produced when 5.8 g of sodium peroxide reacts completely with water at s.t.p.

   (O = 16, Na = 23, V_m at s.t.p. = 22.4 dm^3 mol^{-1})
7. On vigorous heating, zinc burns in oxygen to form a white solid, zinc oxide, according to the following

equation.

$$2Zn(s) + O_2(g) \rightarrow 2ZnO(s)$$

Calculate the mass of zinc oxide produced when 50.0 cm^3 of oxygen completely reacts with zinc at s.t.p.

(O = 16, Zn = 65, V_m at s.t.p. = 22.4 $dm^3\ mol^{-1}$)

8. One of the chemical properties of ammonia gas is that it reacts with acids to form salts. An example is as follows.

$$2NH_3(g) + H_2SO_4(aq) \rightarrow (NH_4)_2SO_4(aq)$$

Calculate the mass of ammonium sulfate that would be produced from the complete reaction of 50.0 cm^3 of a 0.15-mol dm^{-3} sulfuric acid solution with ammonia.

(H = 1, N = 14, O = 16, S = 32)

9. 4.5 g of magnesium is mixed with an excess solution of hydrochloric acid. Calculate the mass of magnesium chloride produced. The equation of reaction is as follows.

$$Mg(s) + 2HCl(aq) \rightarrow MgCl_2(aq) + H_2(g)$$

(Mg = 24, Cl = 35.5)

10. Chlorine gas dissolves in water to form hypochlorous acid. The equation of reaction is as follows.

$$H_2O(l) + Cl_2(g) \rightarrow HClO(aq) + HCl(aq)$$

On exposure to sunlight, the acid decomposes as follows.

$$HClO(aq) \rightarrow 2HCl(aq) + O_2(g)$$

Calculate the maximum volume of oxygen expected from the decomposition of the hypochlorous acid produced when 250 cm^3 of chlorine completely reacts with water at s.t.p.

(V_m = 22.4 $dm^3\ mol^{-1}$)

11. 1.5 mol of hydrogen react completely with nitrogen to form ammonia. What mass of ammonium chloride would be produced if the ammonia produced reacts completely with hydrochloric acid? The equation of reaction is as follows.

$$NH_3(g) + HCl(aq) \rightarrow NH_4Cl(aq)$$

(H = 1, N = 14, Cl = 35.5)

CHAPTER 11: LIMITING REAGENT

Except reactants are mixed together in stoichiometric proportions, the amounts of the products formed in a chemical reaction are determined by the amount of one reactant known as the limiting reagent. In other words, a limiting reagent is the reactant which determines the maximum amounts of the products that can be formed in a chemical reaction.

A limiting reagent is usually the reactant which leads to the production of the least amount of any of the products in a chemical reaction. A reaction stops once all the limiting reagent has been used up.

Example 1

Nitrogen monoxide is reduced to nitrogen by burning in phosphorus. The equation of reaction is as follows.

$$P_4(s) + 10NO(g) \rightarrow 2P_2O_5(s) + 5N_2(g)$$

(a) Determine the limiting reagent when 3.5 mol of phosphorus are mixed with 5.0 mol of nitrogen monoxide.

(b) The amount of nitrogen produced in the above reaction.

Solution

(a) As we stated above, a limiting reagent leads to the production of the least amount of a particular product. All we need to do is to calculate the amount of any product, say N_2, that would be produced by each of the two reactants.

We will begin with P_4. Let the amount of N_2 that will be produced from produced from 3.5 mol of P_4 be n. According to the equation, 1 mol of P4 would produce 5 mol of N_2; hence,

$$\frac{n}{5 \text{ mol}} = \frac{3.5 \text{ mol}}{1 \text{ mol}}$$

$$n \times 1 \text{ mol} = 3.5 \text{ mol} \times 5 \text{ mol}$$

$$n = \frac{3.5 \text{ mol} \times 5 \text{ mol}}{1 \text{ mol}} = 17.5 \text{ mol}$$

The next step is to repeat the procedure for NO. Let the amount of N_2 that would be produced from 5.0 mol of NO be n. According to the equation, 2 mol of NO would produce 1 mol of N_2; hence,

$$\frac{n}{1 \text{ mol}} = \frac{5.0 \text{ mol}}{2 \text{ mol}}$$

$$n \times 2 \text{ mol} = 5.0 \text{ mol} \times 1 \text{ mol}$$

$$n = \frac{5.0 \text{ mol} \times 1 \text{ mol}}{2 \text{ mol}} = 2.5 \text{ mol}$$

Thus, the reactants, NO and P_4, would lead to the production of 2.5 mol and 17.5 mol of N_2, respectively. Since NO will lead to the production of the lesser amount of the product, then it Is the limiting reagent.

We would, of course, arrive at the same answer if we used P_2O_5 as the basis for our calculation. You should be able to verify this for practice.

(b) As stated above, the maximum amount of a product that can be produced in a chemical reaction is that produced by a limiting reagent. Thus, the amount of nitrogen produced is 2.5 mol.

Example 2

One of the properties of metals is that those that are higher up the Activity Series displace those that are lower down the series from solutions of their salts. An example is the following reaction.

$$Cu(s) + 2AgNO_3(aq) \rightarrow Cu(NO_3)_2(aq) + Ag(s)$$

Suppose we mix 75 g of copper with 88 g of silver nitrate, determine:

(a) the limiting reagent;

(b) the mass of any reactant left.

(H = 1, N = 14, O = 16, Cu = 63.5, Ag = 107.9)

Solution

We will use silver, Ag, as the basis for our calculation.

Let the mass of Ag produced from 75 g of Cu be m. According to the equation, 63.5 g of Cu will produce 107.9 g of Ag; hence,

$$m = 75\text{ g}$$
$$107.9\text{ g} = 63.5\text{ g}$$
$$m \times 63.5\text{ g} = 75\text{ g} \times 107.9\text{ g}$$
$$m = \frac{75\text{ g} \times 63.5\text{ g}}{107.9\text{ g}} = 127\text{ g}$$

Next, we will repeat the procedure for the second reactant, $AgNO_3$. Let the mass of Ag produced from 88 g of AgNO3 be m. According to the equation, 339.8 g of $AgNO_3$ would produce 107.9 g of Ag; hence,

$$m = 88\text{ g}$$
$$107.9\text{ g} = 339.8\text{ g}$$
$$m \times 339.8\text{ g} = 88\text{ g} \times 107.9\text{ g}$$
$$m = \frac{88\text{ g} \times 107.9\text{ g}}{339.8\text{ g}} = 27.9\text{ g}$$

Thus, $AgNO_3$ is the limiting reagent, as it leads to the production of the lower amount of the product.

Note that we would arrive at the same answer if we used $Cu(NO_3)_2$ as the basis for our calculation. You should be able to verify this for practice.

(b) Since all of the limiting reagent is used up in a chemical reaction, then the mass of the reactant left is the mass of excess Cu. To obtain this we have to calculate the mass of Cu that would react with $AgNO_3$. Let the mass of Cu required for the complete reaction of 88 g of AgNO3 be m. According to the equation, 63.5 g of Cu requires 339.8 g of AgNO3; hence,

$$m = 88\text{ g}$$

$$63.5\text{ g} = 339.8\text{ g}$$

$$m \times 339.8\text{ g} = 88\text{ g} \times 63.5\text{ g}$$

$$m = \frac{88\text{ g} \times 63.5\text{ g}}{339.8\text{ g}} = 16.4\text{ g}$$

$\therefore$ Mass of reactant left = 75 g – 16.4 g = 59 g

1. One of the chemical properties of nitrogen dioxide is that it oxidizes red-hot metals, while it is itself reduced to nitrogen. An example is the following reaction.

$$4Cu(s) + 2NO_2(g) \rightarrow 4CuO(s) + N_2(g)$$

Determine the limiting reagent when 10 mol of copper is mixed with 15 mol of nitrogen dioxide.

2. Calculate the mass of ammonia that would be produced when 5.0 mol of nitrogen are mixed with 4.5 mol of hydrogen. The equation of reaction is as follows.

$$N_2(g) + 3H_2(g) \rightleftharpoons 2NH_3(g)$$

(H = 1, N = 14)

3. Calcium carbide is produced industrially by heating calcium oxide with coke (carbon) as shown by the following equation.

$$CaO(s) + 3C(s) \rightarrow CaC_2(s) + CO(g)$$

Calculate the mass of any excess reactant when 45 g of calcium oxide is heated with 35 g of coke.

(C = 12, O = 16, Ca = 40)

4. Insoluble carbonates are precipitated by adding sodium carbonate solution to a solution of an appropriate metallic salt. An example is shown by the following equation.

$$CuSO_4(aq) + Na_2CO_3(aq) \rightarrow CuCO_3(s) + Na_2SO_4(aq)$$

Calculate the mass of the salt that would be formed if 150 g of copper(II) sulfate is mixed with 85 g of sodium carbonate.

(C = 12, O = 16, Na = 23, S = 32, Cu = 63.5)

CHAPTER 12: GAS (VOLUME-VOLUME) STOICHIOMETRY

The number of moles of a gas is proportional to its volume. Consequently, in reactions involving only gases, the stoichiometric coefficients can be treated as volumes provided the pressure and temperature remain constant. This assertion is based on Avogadro's law, which states that equal volumes of all gases at the same temperature and pressure contain the same number of molecules. An example of a homogeneous gaseous reaction is the conversion of carbon monoxide to carbon dioxide:

	$2CO(g)$	+	$O_2(g)$	$\rightarrow$	$2CO_2(g)$
Mole	2		1		2
Volume	2		1		2
Mole/volume ratio	2	:	1	:	2

The calculations in gas stoichiometry are performed in much the same way as we did in the previous chapter, except that we can now treat stoichiometric coefficients as volumes.

In gas stoichiometry, the temperature and pressure are assumed constant if no mention is made of the conditions of reaction.

Example 1

Ethene burns in oxygen according to the equation:

$$C_2H_4(aq) + 3O_2(g) \rightarrow 2CO_2(g) + 2H_2O(g)$$

What volume of oxygen is required to react with 50.0 cm^3 of ethene?

Solution

Let the volume of O_2 required to react with 50.0 cm^3 of C_2H_4 be V.

According to the equation, 1 cm^3 of C_2H_4 would produce 3 cm^3 of O_2; hence,

$$V = 50.0\ cm^3$$
$$3\ cm^3 = 1\ cm^3$$
$$V \times 1\ cm^3 = 50.0\ cm^3 \times 3\ cm^3$$

Example 2

Propane burns in air according to the following equation.

$$C_3H_8(g) + 5O_2(g) \rightarrow 3CO_2(g) + 4H_2O(g)$$

Calculate the volume of carbon dioxide that would be produced from the complete burning of 25 cm^3 of propane.

Solution

Let the volume of CO_2 that produced from 25 cm^3 of C_3H_8 be V. According to the equation, 1 cm^3 of C_3H_8 would produce 3 cm^3 of CO_2; hence,

$$V = 25\ cm^3$$
$$3\ cm^3 = 1\ cm^3$$
$$V \times 1\ cm^3 = 25\ cm^3 \times 3\ cm^3$$
$$V = \frac{25\ cm^3 \times 3\ cm^3}{1\ cm^3} = 75\ cm^3$$

Example 3

Butane burns in air according to the following equation.

$$2C_4H_{10}(g) + 13O_2(g) \rightarrow 8CO_2(g) + 10H_2O(g)$$

35 cm^3 of propane is sparked with 150 cm^3 of air. If all the gases are measured at s.t.p. and air contains 21% of air, calculate:

(a) the total volume of all residual gases;

(b) the volume residual gases after the mixture is passed through excess calcium hydroxide solution.

Solution

(a) The total volume, V, of residual gases is the sum of the volumes of gases produced and those of reactants that remain unreacted, i.e.

V = volume of CO_2 + volume of H_2O + volume of unreacted air + volume of excess propane

$$\text{Volume of air} = 150\ cm^3$$

$$\text{Volume of } O_2 = \frac{21}{100} \times 150\ cm^3 = 31.5\ cm^3$$

$$\text{Volume of air remaining} = 150\ cm^3 - 31.5\ cm^3$$
$$= 114.5\ cm^3$$

The reactants are not supplied in stoichiometric proportion; hence, we have to determine the limiting reagent. We will use CO_2 as the basis for our calculation. Let the volume of CO_2 produced from 35 cm^3 of C_4H_{10} be V. According to the equation, 2 cm^3 of C_4H_{10} would produce 8 cm^3 of CO_2; hence,

$$V = 35\ cm^3$$
$$8\ cm^3 = 2\ cm^3$$
$$V \times 2\ cm^3 = 35\ cm^3 \times 8\ cm^3$$
$$V = \frac{35\ cm^3 \times 8 cm^3}{2\ cm^3} = 140\ cm^3$$

We will now repeat the procedure for the second reactant, O_2. Let the volume of CO_2 that would be produced from 31.5 cm^3 of O_2 be V. According to the equation, 13 cm^3 of O_2 would produce 8 cm^3 of CO_2; hence,

$$V = 31.5\ cm3$$
$$8\ cm^3 = 13\ cm^3$$
$$V \times 13\ cm^3 = 31.5\ cm^3 \times 8\ cm^3$$
$$V = \frac{31.5\ cm^3 \times 8 cm^3}{13\ cm^3} = 19.4\ cm^3$$

Thus, O_2 is the limiting reagent. This implies that the volume of CO_2 produced is 19.4 cm^3 and that no volume of O_2 is left unreacted.

The next step is to determine the volume of C_4H_{10} that would react with 31.5 cm^3 of O_2. Let this volume be V. According to the equation, 13 cm^3 of O_2 would react with 2 cm^3 of C_4H_{10}; hence,

$$V = 31.5\ cm^3$$
$$2\ cm^3 = 13\ cm^3$$
$$V \times 13\ cm^3 = 31.5\ cm^3 \times 2\ cm^3$$

$$V = \frac{31.5\ cm^3 \times 2\ cm^3}{13\ cm^3} = 4.85\ cm^3$$

$\therefore$ Volume of excess C_4H_{10} = 35cm^3 – 4.85 cm^3
= 30.15 cm^3

We will now determine the volume of H_2O that would be produced when 4.85 cm^3 of C_4H_{10} reacts completely with 31.5 cm^3 of O_2. Let the volume of H_2O that would be produced from 31.5 cm^3 of O_2 be *V*. According to the equation, 13 cm^3 of O_2 would produce 10 cm^3 of H_2O; hence,

$$V = 31.5\ cm^3$$
$$10\ cm^3 = 13\ cm^3$$
$$V \times 13\ cm^3 = 31.5\ cm^3 \times 10\ cm^3$$

$$V = \frac{31.5\ cm^3 \times 10\ cm^3}{13\ cm^3} = 24.23\ cm^3$$

It would make no difference if we based our calculation on C_4H_{10}. Let the volume of H_2O that would be produced from 4.85 cm^3 of C_4H_{10} be *V*. According to the equation, 2 cm^3 of C_4H_{10} would produce 10 cm^3 of H_2O; hence,

$$V = 4.85\ cm^3$$
$$10\ cm^3 = 2\ cm^3$$
$$V \times 2\ cm^3 = 4.85\ cm^3 \times 10\ cm^3$$

$$V = \frac{4.85\ cm^3 \times 10\ cm^3}{2\ cm^3} = 24.25\ cm^3$$

Note that the slight difference is due to rounding.

The summary of our results is given.

	$2C_4H_{10}(g)$ +	$13O_2(g)$ →	$8CO_2(g)$ +	$10H_2O(g)$
Initial volume (cm^3)	35	31.5	–	–

Final volume (cm^3) 30.15 – 19.4 24.23

∴ $V = 30.15\ cm^3 + 19.4\ cm^3 + 24.23\ cm^3 + 114.5\ cm^3$

$= 188\ cm^3$

(b) CO_2 will react with calcium hydroxide; hence,

∴ $V = 30.15\ cm^3 + 114.5\ cm^3 + 24.23\ cm^3 = 168.9\ cm^3$

Practice problems

1. Carbon monoxide is oxidised to carbon dioxide as follows.

$$2CO(g) + O_2(g) \rightarrow 2CO_2(g)$$

What volume of oxygen is required to react with 50 cm^3 of carbon monoxide?

2. Ammonia reacts with oxygen to produce nitrogen and water as follows.

$$4NH_3(g) + 3O_2(g) \rightarrow 2N_2(g) + 6H_2O(l)$$

What volume of nitrogen would be produced when 150 cm3 of ammonia is sparked with 100 cm3 of oxygen at constant temperature and pressure?

3. 80 cm^3 of nitrogen monoxide is sparked with 150 cm^3 of air. Determine the total volume of the resulting gas mixture if all the gases are measured at s.t.p. Assume that air contains 21% of air. The equation of reaction is as follows.

$$2NO(g) + O_2(g) \rightarrow 2NO_2(g)$$

4. 50 cm^3 of ethyne was sparked with 200 cm^3 of oxygen. Determine the total volume of the resulting gas mixture after it was passed through excess calcium hydroxide solution.

CHAPTER13: PERCENT YIELD

Due to such factors like loss of reactants, side reactions (formation of some other substances that are not shown by the equation of reaction) and the presence of impurities in the reactants, the actual amounts of products obtained from a chemical reaction are usually lower than those obtained from stoichiometric calculations. For example, in the reaction $CaCO_3(s) \rightarrow CaO(s) + CO_2(g)$, carbon monoxide may also be formed, lowering the amounts of the products.

The amount of a product obtained from stoichiometry calculations is called theoretical yield, while that obtained in practice is called actual yield. The percent yield of a reaction is its actual yield expressed as a percentage of its theoretical yield. This given by

$$\text{Percent yield} = \frac{\text{Actual yield}}{\text{Theoretical yield}} \times 100\%$$

Example

Potassium chlorate dissociates as follows.

$$2KClO_3(s) \rightarrow 2KCl(s) + 3O_2(g)$$

Determine the percent yield of the reaction if the actual volume of oxygen obtained from the decomposition of 1.5 g of potassium chlorate is 300 cm^3 at s.t.p.

(O = 16, K = 39, Cl = 35.5, V_m at s.t.p. = 22.4 $dm^3\ mol^{-1}$)

Solution

We will apply the relation

$$\text{Percent yield} = \frac{\text{Actual yield}}{\text{Theoretical yield}} \times 100\%$$

The theoretical yield is obtained from stoichiometric calculation. The very first step is to calculate the theoretical number of moles of O_2 produced. The number of moles of $KClO_3$ supplied is obtained as follows.

m = 1.5 g

$M = \{(39 \times 1) + (35.5 \times 1) + (16 \times 3)\}\ \text{g mol}^{-1}$

$= 122.5\ \text{g mol}^{-1}$

n = ?

Substituting we have

$$n = \frac{1.5\ \text{g} \times 1\ \text{mol}}{122.5\ \text{g}} = 0.01224\ \text{mol}$$

Let the number of moles of O_2 produced from 0.01224 mol of $KClO_3$ be n. According to the equation, 2 mol of $KClO_3$ would produce 3 mol of O_2; hence,

n = 0.01224 mol

2 mol = 3 mol

$n \times 3\ \text{mol} = 0.01224\ \text{mol} \times 2\ \text{mol}$

At s.t.p. we have

$$n = \frac{V}{22.4\ \text{dm}^3\ \text{mol}^{-1}}$$

$\therefore$ $V = n \times 22.4\ \text{dm}^3\ \text{mol}^{-1}$

V = ?

$$V = 0.01836\ \text{mol} \times \frac{22.4\ \text{dm}^3}{1\ \text{mol}} = 0.411\ \text{dm}^3$$

Actual yield = 300 cm^3 = 0.30 dm^3

Percent yield = ?

$$\text{Percent yield} = \frac{0.30\ \text{dm}^3}{0.411\ \text{dm}^3} \times 100\% = 73\%$$

Practice problems

1. 2.5 g of copper were obtained when 4.5 g copper(II) oxide was

mixed with excess carbon monoxide. Determine the percent yield of copper. The equation of reaction is as follows.

$$CuO(s) + CO(g) \rightarrow Cu(s) + CO_2(g)$$

(O = 16, Cu = 63.5)

2. Insoluble bases are prepared by precipitation. An example is the reaction between copper(II) sulfate and potassium hydroxide to yield potassium sulfate and copper(II) hydroxide. Given that the mass of copper(II) hydroxide obtained when 50.0 cm^3 of a 0.15-mol dm^{-3} solution of potassium hydroxide was mixed with 50.0 cm^3 of a 0.20-mol dm^3 solution of copper(II) sulfate was 0.25 g, determine the percent yield of the reaction. The equation of reaction is as follows.

$$CuSO_4(aq) + 2KOH(aq) \rightarrow K_2SO_4(aq) + Cu(OH)_2(s)$$

(H = 1, O = 16, Cu = 63.5)

ANSWERS

Chapter 1

1. AgCl
2. $Ca(OH)_2$
3. $(NH_4)_2SO_4$
4. $FeCl_2$
5. $Mg(NO_3)_2$

Chapter 2

1. 28 g mol^{-1}
2. 56 g mol^{-1}
3. 17 g mol^{-1}
4. 44 g mol^{-1}
5. 111 g mol^{-1}
6. 100 g mol^{-1}
7. 62 g mol^{-1}
8. 74 g mol^{-1}
9. 80 g mol^{-1}
10. 278 g mol^{-1}

Chapter 3

1. C = 0.64 g; O = 0.86 g
2. C= 4.3 g; H = 0.71 g
3. C = 2.6 g; H = 0.65 g, O = 1.8 g
4. (a) N = 82.4%, H = 17.6%
   (b) C = 79.9%, O = 20.1%
   (c) Cu = 88.8%, O = 11.2%
   (d) Ca = 40%, C = 12%, O = 48%
   (e) H = 2.0%, S = 32.7%, O = 65.3%
   (f) Ca = 54.1% O = 43.2%, H = 2.7%

(g) N = 21.2%, H = 6.1%, S = 24.2%, O = 48.5%

Chapter 4

1. (a) 0.016 mol
   (b) 0.00071 mol
   (c) 0.012 mol
2. (a) 5.5 g
   (b) 1.1 g (c) 133.8 g
3. (a) 74 g mol^{-1}
   (b) 2 g mol^{-1}
   (c) 36.5 g mol^{-1}
4. $Na2CO_3.H_2O$

Chapter 5

1. (a) 7.56 mol (b) 0.000498 mol (c) 1.51 mol
2. (a) 1.5×10^{24}
   (b) 6.6×10^{22}
   (c) 3.0×10^{24}
3. (a) 3.65 g
   (b) 4.93 g
   (c) 45. 2 g
4. (a) 4 g mol^{-1}
   (b) 65 g mol^{-1}
   (c) 16 g mol^{-1}

Chapter 6

1. (a) 0.00028 mol

   (b) 0.038 mol

   (c) 0.0020 mol
2. (a) 11 dm^3
   (b) 0.033 dm^3
   (c) 0.65 dm^3
3. (a) 1.0 mol dm^{-3}

(b) 6.0 mol dm^{-3}
(c) 0.66 mol dm^{-3}
4. (a) 9.0×10^{22}
(b) 5.0×10^{22}
(c) 1.7×10^{22}
5. (a) 106 g mol^{-1}
(b) 40 g mol^{-1}
(c) 98 g mol^{-1}

Chapter 7

1. (a) 0.025 mol
(b) 0.011 mol
(c) 0.067 mol
2. (a) 56 dm^3
(b) 3.4 dm^3
(c) 28 dm^3
3. (a) 3.3×10^{24}
(b) 4.0×10^{21}
(c) 6.7×10^{21}
4. (a) 0.9 g
(b) 3.1 g
(c) 0.83 g
5. (a) 28 g mol^{-1}
(b) 34 g mol^{-1}
(c) 16 g mol^{-1}

Chapter 8

1. (a) C_5H_8
(b) C_2H_5
(c) $CaCO_3$
(d) C_2H_5COOH
(e) $C_4H_5N_2O$
2. (a) CH_3
(b) CO_2
(c) C_2H_6O

(d) $CuSO_4.5H_2O$
(e) $C_9H_8O_2$
3. (a) $C_{10}H_{22}$
(b) $C_2H_2O_4$
(c) H_2CO_2
(d) C_3H_6O
(e) $C_6H_{12}O_6$
4. C_3H_7COOH, propanoic acid
5. (a) 276 g mol^{-1}
(b) C_5H_9
(c) $C_{20}H_{36}$
6. (a) 536 g mol^{-1}
(b) C = 89.6%, H = 10.4%
(c) C_5H_7
(d) $C_{40}H_{56}$

Chapter 9

1. $2H_2O_2(g) \rightarrow 2H_2O(l) + O_2(g)$
2. $CH_4(g) + 2O_2(g) \rightarrow CO_2(g) + 2H_2O(g)$
3. Balanced as written
4. Balanced as written
5. $2Na(s) + 2H_2O(l) \rightarrow NaOH(aq) + H_2(g)$
6. $C_5H_{12}(g) + 8O_2(g) \rightarrow 5CO_2(g) + 6H_2O(g)$
7. $3Ca(OH)_2(aq) + H_3PO_4(aq) \rightarrow Ca_3(PO_4)_2(aq) + H_2O(l)$
8. $2AgI(aq) + Na_2S(s) \rightarrow Ag_2S(s) + 2NaI(aq)$
9. Balanced as written
10. Balanced as written

Chapter 10

1. 8.3 mol
2. 6.0 mol
3. 66.9 g
4. 186.3 g
5. 833 cm^3
6. 0.18 g

7. 0.99 g
8. 250 cm^3
9. 17.8 g
10. 53.5 g

Chapter 11

1. Cu
2. H_2
3. 9.7 g of C
4. 113.9 g

Chapter 12

1. 25 cm^3
2. 66.6 cm^3
3. 250 cm^3
4. 225 cm^3

Chapter 13

1. 69.6%
2. 68.3%

APPENDICES

Some scientific constants

Constant	Symbol	Value
Gravitational constant	G	6.670×10^{-11} N m^2 kg^{-2}
Velocity of light	c	2.9979×108 m s^{-1}
Atomic mass unit	amu	1.6605×10^{-27} kg
Avogadro's number	N_A	6.0221×10^{23} mol^{-1}
Boltzmann's constant	k	1.3807×10^{-23} J K^{-1}
Mass of electron	m_e	9.109×10^{-31} kg
Mass of neutron	m_n	1.6726×10^{-27} kg
Mass of proton	m_p	1.6726×10^{-27} kg
Faraday's constant	F	9.64870×10^4 C mol^{-1}
Electron charge	e	9.1096×10^{-31} kg
Gas constant	R	8.314 J K^{-1} mol^{-1}
Molar volume of gases at s.t.p.	V_m	22.4136 dm^{-3} mol^{-1}
Electron radius		2.8177×10^{-15} m
Rydberg's constant	R_H	1.0974 m^{-1}
Acceleration due to gravity	g	9.8066 m s^{-2}

SI Base units

Quantity	Unit	Symbol
Length	Metre	m
Mass	Kilogram	kg
Time	Second	s
Electric current	Ampere	A
Temperature	Kelvin	K

Amount of substance Mole mol

Multiples of SI units

Prefix	Symbol	Value
kilo-	k	10^3
mega-	m	10^6
Giga-	G	10^9
Tera-	T	10^{12}

Sub-multiples of SI units

Prefix	Symbol	Value
deci-	d	10^{-1}
centi-	c	10^{-2}
milli-	m	10^{-3}
micro-	μ	10^{-6}
nano-	n	10^{-9}
pico-	p	10^{-12}
femto-	f	10^{-15}
atto-	a	10^{-18}

Nobel prize winners in chemistry

1901: Jacobus H. van 't Hoff
1902: Hermann Emil Fischer
1903: Svante Arrhenius
1904: Sir William Ramsay
1905: Adolf von Baeyer 1906: Henri Moissan
1907: Eduard Buchner
1908: Ernest Rutherford
1909: Wilhelm Ostwald
1910: Otto Wallach
1911: Marie Curie
1912: Victor Grignard and Paul Sabatier 1913: Alfred

Werner
1914-1918: Prize not awarded 1919: Fritz Haber
1920: Walther Nernst
1921: Frederick Soddy 1922: Francis W. Aston 1923: Fritz Pregl
1924: Not awarded
1925: Richard Zsigmondy
1926: Theodor Svedberg
1927: Heinrich Wieland
1928: Adolf Windaus
1929: Arthur Harden and Hans von Euler-Chelpin 1930: Hans Fischer
1931: Carl Bosch and Friedrich Bergius 1932: Irving Langmuir
1933: Paul Dirac and Erwin Schrödinger 1934: Harold C. Urey
1935: Frédéric Joliot and Irène Joliot-Curie 1936: Peter Debye
1937: Norman Haworth and Paul Karrer 1938: Richard Kuhn
1939: Adolf Butenandt and Leopold Ruzicka 1940-1942: Prize not awarded
1943: George de Hevesy 1944: Otto Hahn
1945: Artturi Virtanen
1946: James B. Sumner, John H. Northrop and Wendell M. Stanley 1947: Sir Robert Robinson
1948: Arne Tiselius
1949: William F. Giauque
1950: Otto Diels and Kurt Alder
1951: Edwin M. McMillan and Glenn T. Seaborg
1952: Archer John Porter Martin and Richard L. M. Synge 1953: Hermann Staudinger
1954: Linus Pauling
1955: Vincent du Vigneaud
1956: Sir Cyril Hinshelwood and Nikolay Semyonov 1957: Lord Todd
1958: Frederick Sanger

1959: Jaroslav Heyrovsky 1960: Willard F. Libby 1961: Melvin Calvin
1962: Max F. Perutz and John C. Kendrew 1963: Karl Ziegler and Giulio Natta
1964: Dorothy Crowfoot Hodgkin 1965: Robert Burns Woodward 1966: Robert S. Mulliken
1967: Manfred Eigen and Ronald G. W. Norrish and George Porter 1968: Lars Onsager
1969: Derek H. R. Barton and Odd Hassel 1970: Luis Federico Leloir
1971: Gerhard Herzberg
1972: Christian B. Anfinsen, Stanford Moore, and William H. Stein 1973: Ernst Otto Fischer and Geoffrey Wilkinson
1974: Paul J. Flory
1975: John W. Cornforth and Vladimir Prelog 1976: William Lipscomb
1977: Ilya Prigogine 1978: Peter D. Mitchell
1979: Herbert C. Brown and Georg Wittig
1980: Paul Berg, Walter Gilbert, and Frederick Sanger 1981:
Kenichi Fukui and Roald Hoffmann
1982: Aaron Klug
1983: Henry Taube
1984: Robert Bruce Merrifield
1985: Herbert A. Hauptman and Jerome Karle
1986: Dudley R. Herschbach, Yuan T. Lee, and John C. Polanyi
1987: Donald J. Cram, Jean-Marie Lehn, and Charles J. Pedersen f
1988: Johann Deisenhofer, Robert Huber, and Hartmut Michel 1989: Sidney Altman and Thomas R. Cech
1990: Elias James Corey 1991: Richard R. Ernst 1992: Rudolph A. Marcus
1993: Kary B. Mullis and Michael Smith 1994: George A. Olah
1995: Paul J. Crutzen, Mario J. Molina, and F. Sherwood

Rowland
1996: Robert F. Curl Jr., Sir Harold W. Kroto, and Richard E. Smalley
1997: Paul D. Boyer and John E. Walker
1998: Walter Kohn and John A. Pople 1999: Ahmed Zewail
2000: Alan J. Heeger, Alan G. MacDiarmid, and Hideki Shirakawa
2001: William S. Knowles and Ryoji Noyori
2002: John B. Fenn and Koichi Tanaka
2003: Peter Agre and Roderick MacKinnon
2004: Aaron Ciechanover, Avram Hershko, and Irwin Rose
2005: Yves Chauvin, Robert H. Grubbs, and Richard R. Schrock
2006: Roger D. Kornberg
2007: Gerhard Ertl
2008: Osamu Shimomura , Martin Chalfie, and Roger Y. Tsien
2009: VenkatramanRamakrishnan,ThomasA.Steitz, and Ada E. Yonath
2010: Richard F. Heck, Ei-ichi Negishi, and Akira Suzuki
2011: Dan Shechtman
2012: Robert J. Lefkowitz and Brian K. Kobilka
2013: Martin Karplus, Michael Levitt, and Arieh Warshel
2014: Eric Betzig , Stefan W. Hell, and William E. Moerner
2015: Tomas Lindahl, Paul Modrich, and Aziz Sancar
2016: Jean-Pierre Sauvage, Sir J. Fraser Stoddart , and Bernard L. Feringa
2017: Jacques Dubochet, Joachim Frank, and Richard Henderson
2018: Frances H. Arnold, George P. Smith, and Sir Gregory P. Winter

2019: John B. Goodenough, M. Stanley Whittingham, and Akira Yoshino
2020: Emmanuelle Charpentier and Jennifer A. Doudna
2021: Benjamin List and David W.C. MacMillan
2022: Carolyn R. Bertozzi, Morten meldal, and K. Barry Sharpless

www.ingramcontent.com/pod-product-compliance
Lightning Source LLC
Chambersburg PA
CBHW071943210526
45479CB00002B/799
* 9 7 9 8 3 3 2 7 6 9 6 1 0 *